MW00625775

34170091508532

THE MOURNER'S BESTIARY

EIREN CAFFALL

Copyright © 2024 by Eiren Caffall

Row House Publishing recognizes that the power of
justice-centered storytelling isn't a phenomenon; it is essential for
progress. We believe in equity and activism, and that books—and
the culture around them—have the potential to transform the
universal conversation around what it means to be human.

Part of honoring that conversation is protecting the intellectual
property of authors. Reproducing any portion of this book (except for
the use of short quotations for review purposes) without the expressed
written permission of the copyright owner(s) is strictly prohibited.
Submit all requests for usage to rights@rowhousepublishing.com.

Thank you for being an important part of the conversation
and holding sacred the critical work of our authors.

Library of Congress Cataloging-in-Publication Data Available Upon Request

ISBN 978-1-955905-58-9 (HC)
ISBN 978-1-955905-59-6 (eBook)

Printed in the United States
Distributed by Simon & Schuster
First edition
10 9 8 7 6 5 4 3 2 1

For Nancy and Charlie

This animal lifts its back out of the open sea above the watery waves, and then it anchors itself in one place; and on its back, what with the shingle of the ocean drawn there by the gales, a level lawn gets made, and bushes begin to grow there. Sailing ships that happen to be going that way take it to be an island, and land on it. Then they make themselves a fireplace. But the Whale, feeling the hotness of the fire, suddenly plunges down into the depths of the deep, and pulls down the anchored ship with it into the profound.

—T. H. WHITE, *THE BOOK OF BEASTS*

THE MOURNER'S
BESTIARY

Contents

MONHEGAN ISLAND, 2014

I say that it touches a man that his blood
is sea water and his tears are salt, that the
seed of his loins is scarcely different than
the same cells in seaweed, and that of
stuff like his bones are coral made.

—DONALD CULROSS PEATTIE,
AN ALMANAC FOR MODERNS

Longfin Inshore Squid

(*Doryteuthis pealeii*)

Dex and I were alone, twelve miles out to sea, nor'easter rolling in over the Gulf of Maine, in the world's fastest-warming marine ecosystem. Dex was nine and I was forty-three and I was supposed to be dead. On the night of the storm, we went to the Monhegan Island dock and met a boy who had just fished a small squid from the harbor waters. The boy had the squid in a bucket, waiting to send her back to sea once he was done staring at her. He scooped her from the seawater and held her out to us. She was a longfin squid, newly moved north to those warming waters. A small female subadult, she pulsated in his hand, nearly translucent, silvery white, red eyes, iridescent, her inner workings revealed—a slender bag of hydrostatic wonder and life, so different from us.

I was a single mother, diagnosed young with an incurable genetic kidney disease that killed most of my family before they reached fifty. I was told I wouldn't even make that. I didn't know if Dex had inherited what I'd grown up calling the Caffall Curse. I was raised by the expendable ill alongside threatened sea creatures in collapsing oceans—tender beings caught in extinction. My whole family was like them, brave and chaotic, poisoned by things outside of their control. I inherited that as much as the disease. Deciding to have Dex, the last of our line, was an act of faith in our futures and bodies, in the world itself; but that faith was costing me.

On the drive to the island, we listened to Paul Simon's *Graceland*. *My traveling companion is nine years old/he is the child of my first marriage*, we sang, the words completely true, as we drove Route 1, Southern Maine retreating, the Midcoast and the ferry ahead. We stood on the bow of the mail boat, wet from the swells on Muscongus Bay, watching a harbor porpoise pod and diving storm petrels. We saw, rising like a whale back, Monhegan, which all winter we called simply "The Island," a narrow strip of land—three-quarters of a mile wide, one and three-quarters of a mile long, with almost four hundred acres of protected wildlands—home to the tide pools and the shipwreck, home to our wonder and dread.

From the dock, we walked the dirt road over the hill, our feet kicking up the same dust as the other tourists. One of three pickup trucks that were allowed on the island delivered our bags to the hotel room. "Can we walk to Lobster Cove?" Dex asked as we changed to drier clothes. The next day we'd move to a cottage, the first one we'd ever rented there, the first time we'd been able to stay more than a few days. I'd worked extra jobs and extra hours to pay for it. I'd paid in advance before the work dried up and planned the research trip for months.

"I think so," I said. "I'm pretty tired, kiddo."

"We always go the first night."

"You're right, we do."

"And then we can go to the dock, and then get a pizza."

"OK. But bring your slicker. The forecast says rain. Maybe a big storm."

"I love it here when it storms."

We took the road from our hotel up the hill to the trail to Lobster Cove, where a narrow single-track snaked from the domestic peace of cottages, down into a tunnel of young trees, emerging into a wide-open vista at the top of a bluff. From the bluff, the land below stretched out in magic-hour light, a broad apron of meadows meeting the sea. The rocky beach of Lobster Cove was full of high tide, black with gathered seaweed, stinking of salinity and marine rot. Beyond the inlet was the open ocean, unobstructed all the way to Spain. The rocky shoreline was bordered by scrub pine and laced with footpaths, a gray tumble of outcrops topped with the orange, oxidized metal of the wreck of the *D. T. Sheridan*, its broad lozenge of hull perched like a sentinel. We walked through it—labyrinth, prayer, memory—scrambling down to the tide pools. The tide was coming in; waves rolled up, pouring themselves into the pools, pulsing along their borders of sea wrack, dulse, and Irish moss.

"Don't go too close, those waves are big."

"I won't." Dex put his feet in the water.

"Careful of the barnacles." I sat down on a boulder near the pool.

"I know, Mom."

I was exhausted, my back sore, my ankle swollen with edema from the drive. The warmth of the rocks came through the seat of my jeans, warmed my skin, my bones, the new heaviness of my body. I fell asleep, waking as my chin hit my chest.

There was collapse in our family, my divorce still an open wound. There was collapse in the water; out beyond the harbor, the Gulf warmed in its benthic depths, losing animals, losing balance. There was collapse in me, invisible to the naked eye as the losses of the ocean. My illness was, by that afternoon at the tide pools, a generational project that went back for at least one hundred and fifty years in my family. I was born in a flooding body, on a flooding planet, in a flooding family—a mourner, a drowner.

Polycystic kidney disease, PKD, the genetic illness I'd inherited, is a mutation—most commonly expressed in adulthood, though existing in a juvenile form—that causes fluid-filled cysts to multiply in the kidneys until they grow and swell, overwhelming all healthy tissue. Afflicted kidneys look as if they are sculpted out of rosy bubble wrap and can be enormous; the largest set of PKD kidneys ever recorded reached a total weight of seventy-seven pounds. The illness can be painful, or it can be quiet. It can bring storms of brain aneurysms, killing the patient before they know they are sick, or start kidney failure so slow that their family is witness to a steady decline. There are forms of PKD—one of the most common genetic illnesses in the world—that linger, hesitating to express themselves in a carrier until that person's seventies. But for most patients, the disease is invisible until it isn't, waiting until a person is childbearing age to begin to flood the body, too much water, too many chemicals in the blood, slowly and relentlessly, a rising sea under the skin.

I have a friend who shares my disease. When she was diagnosed, her nephrologist told her each cyst operates as its own ecosystem, a tide pool—with specific pathogens and bacteria, unfiltered toxic sludge from the blood making each bubble its own world. Each body with PKD is its own ecosystem of water and consequences. Genetic illness echoes back and forth along the lines of time, along cells, along strands of DNA—a powerful whirlpool that can pull you down.

The nephrologist who diagnosed me told me only that I'd die young. My family, four generations into the disease, had given up on survival. I gave up on a safe future, became a single musician living in Chicago, spending my hours in bars with sticky floors covered in cigarette butts, watching music, ecstatic with the sound vibrating my ribs, chasing after feeling and transcendence in the body-high of late nights and voices braiding into my skin and leaving me breathless and less alone. I wanted my body to be something else—a channel for that feeling, an instrument of good until I couldn't use it anymore. But even as I sang in rooms full of people, I hid from what had brought me there, language and song obscuring all the loss. Once I got a job with health insurance, I went to a nephrologist for the first time since I was diagnosed.

"Do you want kids?" Dr. K said.

"Yeah. I think I want kids. Eventually."

"I know how to end PKD in one generation." Dr. K smiled, raising one eyebrow.

"You do?"

"People like you should never have children."

My father's siblings had all made the decision not to have biological children, wanting the curse to end with them. My father thought he was free of the disease when I was born, the plague missing our house, learning only the next year that he had it too. My parents never had any more kids. I know people who wrestle with the question of whether or not to make new generations as the planet warms and collapses. The activists who work around ideas of reproduction and climate are clear—the fact that we ask ourselves whether it is safe to have children in the crisis of environmental collapse is the problem itself. We are told that our individual sacrifices are the last chance to end collapse. The Caffalls told themselves, as that nephrologist told me, that our individual sacrifices were the last chance to end generations of mourning. These are lies.

It is easy to tell people that their existence, need, joy, vulnerability, humanity is the driver of collapse. It is harder for the world to change to protect vulnerable lives, human and nonhuman. It takes believing that a life like mine matters, or the life of a plankter or a littleneck clam. It took my father dying to convince me I could want something so dangerous—a child of my own, another one of us, the broken. I married badly to make it happen. I put my health at risk.

"I think there are fewer periwinkles," Dex said, holding up the shell of a snail for me to see.

"Any urchins this year?"

"Not yet. I'm going to see more snails." He bent back over the pool.

Babylonian clay tablets exist that record in great detail the diseases of the kidney, some of which might be PKD. There is familiarity in the language of the neo-Assyrian doctor who wrote, "If a man suffers . . . and it strikes his kidney. . . that man is being pursued by the hand of a ghost." Early clinicians of PKD called the cysts *the teardrops of the kidneys*—a mourner's disease from the start.

"Come here, Mom, see these snails I found." Dex beckoned me closer. I stood up and walked toward him.

Squatting next to the water, I touched the pile he'd collected, many beginning the long walk back to the pools from the dry rock where they suddenly found themselves. "That's a lot."

"But not as many as before."

I was a mourner that night, kin to the broken ocean since birth, the daughter of a drowning man and a water scientist, one of a line of family obsessed with the sea and its creatures, mother to the same. We think we know mourning because we know grief. We think we know grief because we have been taught that it is a linear walk of five stages. But neither grief, the private cost of loss, nor mourning, its public cousin of practice and connection, are reducible to five stages, walked through in a straight line and completed. It is instead a spiral, where we come back to each stage again and again.

On the way to the island, as I drove us from Lake Michigan to Monhegan, Dex begged to hear his favorite audiobook, *Twenty Thousand Leagues Under the Sea,* over and over again—Captain Nemo and his *Nautilus* on the car speakers. *Mobilis in mobili,* goes the bad Latin of Nemo's motto—supposedly *moving in a moving element,* but more closely, *moving in a moving thing, mobile in mobile.* Like mourning, always moving, like water, changing all the time.

Dex handed me a snail in a pink-and-yellow swirling shell, and I wondered how I could hold on to all the stories of my body. The submarine in the novel was named for a Pacific Ocean mollusk, a creature swimming backward, seeing always where it started. The chambers of a nautilus are each built on the last, their structure a spiral order. That chambering was ancient, deep in the fossil record as a body plan—a survival plan. Mourning my well body, I was in the stages of grief. I wanted to walk each step in that promised straight line, walling them off as I went. But the whirlpool I had entered wasn't a simple walk of five steps, completed and contained, but that spiral that brings the mourner through the stages over a lifetime.

I stared at Dex and his snails and the warming water. He was focused on fish just as I had been at his age. *We're making a bestiary,* I thought, *a list of fish that will teach me how to live.*

When first popular in medieval Europe, bestiaries were moral teaching tools, spiritual instruction superimposed over the bodies of animals. They became the earliest encyclopedias, the first natural history manuals, explaining animals to people and people to themselves. They were safaris, truths, fictions, scriptural, folkloric. The unicorn was Christ, she was chastity. The whale was a mountain, an island, the enormity and surprise of the devil. I knew, sitting there on the hard rocks of the island, exhausted, in pain and isolated, that I needed the stories of water animals to understand the loss in my family and in the families of creatures in the water around us. *What is the life of that snail*, I wondered, beginning the first entry in my head.

"I should help these guys back into the pool," Dex said.

"Want a hand?"

"OK."

"We can check on them again when we come back tomorrow," I said. "Let's go to the dock before it gets dark."

We climbed back up the hill, passed the dainty green-and-white sign nailed to a tree stump that said *Town* with an arrow pointing up, passed brambles of blackberries twined around the limbs of serviceberry and young ash. Somewhere, on the lawn of a wealthy homeowner, someone was mowing, and the smell of cut grass merged with the oncoming rain and the acrid clean salt, a smell my mother trained me to search for on car trips.

"Roll down your window," she'd yell to the back seat of the station wagon, "there's a salt marsh!" Or "You can smell the tide," she'd say as we came to the Long Island Sound. The Sound was her water, the place where my grandmother, Mac, had tried to teach me to swim, the place where I'd first seen what it would take to swim through the rising sea. I was afraid I hadn't learned to swim through those waters, in my body or the world. I was afraid I hadn't taught Dex to swim them either.

I could see Dex's bare feet on the path ahead, the bob of his long hair as he jumped from rock to rock. The year was wet, and the dirt around the stones in the path was muddy, the rocks slick.

"It's all the same," Dex said. "Like we left it last year."

But it wasn't like last year. Work had dried up. We were on Medicaid. Friends sent me cash cards to pay for groceries. I fed Dex before myself. I scrambled for work to take with me to the island. On the drive out, I skipped seeing my mother. Her love for me swung between grand gestures of care and the abyss of neglect, between bravery and fear, punctuated by emotional heavy weather I learned to avoid. It felt like Dex's father, Jonah, had a radar for years when I was low. But there was something worse—I was getting sick.

"Are you sure you're OK to drive to Maine alone?" my friend Amanda asked.

"No," I admitted.

"And straight there? That's twenty hours."

"I know, but I can't afford a hotel to stop. And Dex . . ."

"Dex needs Monhegan?"

"Yeah."

"Fair."

"And so do I. It's already paid for. And I have research to do."

"But the pain is worse, right?"

"Yeah. But that's probably just stress."

"When did you see your nephrologist last?"

"I don't know. My old doctor doesn't take Medicaid."

"Will you do it when you get back?"

"Yeah. OK."

Twilight was coming, and I remembered there was no doctor on the island, no emergency response boat. The village and its sixty year-round residents ringed the narrow working harbor cosseted by its sister island, treeless Manana. The village was car-free, apart from a few working trucks, bisected by dirt roads. The marsh and meadow at the center were the source of the island's limited fresh water—a deep basin of green surrounded by clapboard houses, those houses giving way to wilderness, one of the nation's oldest private land conservancies, the highest cliffs in Maine, the lighthouse. Helicopters landed on Lighthouse Hill in case of medical emergency, but only in good weather. Monhegan's green hills and glowing sea were a risky place to come. I tried not to know that in my body. Instead, I reminded myself that this island and this sea did not belong to me. I was not a resident. I was a pilgrim.

Advancing clouds hid the sunset. We crossed through the village, over the hill to the dock where we'd landed that afternoon. When ferries leave for the mainland, passengers throw bouquets of wildflowers from the stern to ensure a safe return; their friends leap from the dock into the freezing Atlantic, so cold you forget it is warming, to seal the bargain. When we got there, the boy and the squid were alone on the wooden dock in the twilight, in that liminal place where journeys end and begin.

His longfin squid was not under threat in a warming world; warmth made them thrive. They were abundant enough to provide a potentially stable food fishery. They were the stars of important medical research. Humans and modern cephalopods share a common ancestor and a key innovation. Ancient cephalopods pioneered the neuron, that single genetic adaptation that allows creatures to think and feel. Human and squid neurons share structural features, including how they translate impulses in the brain into action. Neurons have three parts—the cell body, the dendrites, and the axon. The axon connects the impulses of the dendrites, through the cell body, to its destination. When the neuron tells you to run from danger, it is the axon that delivers the news to your feet. Longfin inshore squid have one of the largest neural axons relative to body size of any creature. While a single impossibly slim human axon cannot be removed from the brain for study, the squid axon can be removed and placed under a microscope to track how electrical impulses move. Discoveries from the study of longfin axons have been awarded Nobel prizes, making possible research that may lead to cures for Alzheimer's and Parkinson's disease. Without the longfin inshore squid, we would not understand how messages of emotion flow through the human body. The waters around us that night held squid—messengers of emotion, messengers of time running out.

Jonah was expecting a call that night, interrupting my time alone with Dex. Our divorce was high conflict. It had been an act of hope for me to have Dex. Jonah knew about the risks of my disease going into the marriage and the pregnancy. I thought he'd protect us both from the damage of PKD, married or not. After we split, furious, he battled me fiercely over the one thing I feared losing the most: time.

My phone pinged. I pulled it out of my pocket.

Eiren?! It is call time! Jonah wrote.

Standing on the dock, I felt sweat on the back of my neck, a prickle of electricity itching against my necklace, fear pulsing down my axons to flood my body. My heart raced in a wave of vertigo and nausea, picturing Jonah waiting by his phone in Chicago, early for the arranged time, angry with me as he always was.

Where are you!?! Jonah wrote.

"We have to go call Daddy," I told Dex as I watched our cell coverage wink out, turning us toward the hotel and better reception.

"Thank you for sharing your squid," Dex said to the boy, who threw his head to the side in a curt farewell, buzz cut standing on end in the wind, and shrugged a Yankee shrug as he tossed the animal back into the harbor. We climbed from the dock, crested the hill to see the island below us. A low fog spread over the marsh, the invasive purple loosestrife and cattails dulled in the twilight. It began to rain, dark came down, the late July night turning cool like autumn. Dex dawdled in the air, the ocean smell, the twilight. I rushed our steps, exhausted, sick, the phone a lead weight in my pocket.

We turned down the hill and under the ancient apple tree at the island's central crossroad. Everything was in my head—another text message coming in, dizziness, sweat creeping toward my hairline. In the oncoming dark, I tripped on a rock, twisted my ankle, and hit the ground hard, chest first.

I sat up. Dex stood next to me, blinking in the rain. A pair of tourists stopped and looked over at us. "I'm fine!" I sang as consciousness left.

Later, Dex would tell me as he lay next to me in a hospital bed, "Mom, you were foaming at the mouth."

Unconscious, I had a vision of my body spinning in circles in a car crash, the vertigo overwhelming. I was surrounded by a ring of protection, the faces of the dead singing the Shaker hymn my father had taught me while we walked on the shore of the Long Island Sound when I was nine, *'Tis a gift to come down where you are to be, and when you have found the place just right, you will be in the valley of love and delight.*

"Oh," I said to the faces, and the ring of family dissolved into a circle of strangers, wind speeding the clouds over us, rain in my mouth. Dex was far on the edge of the people, *The Boy Who Is Where Others Are Not*, who'd searched out bees alone in the playground's privet hedge.

"You OK?" A woman's voice.

I swiveled my rag doll head around to find her. As a child, I couldn't sleep. To help me, my mother would play a game. I'd lie on the floor in the twilight while she gently shook each of my limbs until they softened, singing, *My name is Flopsy Flora/a dolly full of rags*, while I tried to relax under her touch, *My arms go flop, and my legs do plop*, tried to give in to the black vertigo of sleep, *and my head just wigs and wags*, tried to rely on the promise that she would keep me safe, a promise I knew wasn't real. She was so beautiful then, before the years of walking dying rivers, before the years of caring for my bohemian father and his flooding body and his reckless choices.

"You OK? You with us?" the stranger asked. She had brown hair, a rain hood pulled up over it. "The emergency team is here. Do you know what happened to you?"

"No."

"Does anyone else?"

"I'm alone."

"Do you have a seizure disorder? Those people saw, said it was a seizure?"

"No. I have kidney disease."

"OK. Well, with a suspected seizure we have to get you off island."

"I'm alone," I said again. I couldn't understand what had happened. My body had failed. I'd been waiting for it to fail—waiting since I was diagnosed at twenty-two, through my father's death, through having Dex, through the divorce and now it had. I looked at the stranger with the dark eyes, and I couldn't begin to explain. "We're alone, and I have kidney disease and my kid can't be away from me."

She blinked at the fear in my voice. "Which one is he?"

I pointed to him, standing isolated in the rain. "We can't be separated," I said. "Can't."

"I got it." She was solemn, a deep well of feeling passed between us. Maybe she was a mother too. "The helicopter can't come—the nor'easter. Fifteen-foot seas. The Coast Guard is sending a cutter." And the sentence held in it all the fear I'd been trying to avoid, a fear that traveled from my brain, through axons sending messages of terror to my limbs, my heart.

There was more that came after that sentence. There was an aftermath. There was diagnosis and discovery. But first, there was the machinery of crisis. The woman sent a young man to our hotel for some things. Later I would learn that she was Jen, a lobsterman's wife, walking her children to school at the one-room Monhegan schoolhouse in the sleeting winters and the rough, bright spring. She was a researcher in college and graduate school, studying the marine ecosystem of the Gulf, and came to Monhegan to work on a project, met the lobsterman, fell in love, shifted her life to the village and the water, and lived there, volunteering for the fire department and for the first aid team. The Coast Guard arrived. Jen walked us over the hill. She held Dex's hand, she held mine. At the dock, she told me the Coast Guard didn't want to let Dex in the boat with me. She said, "I told them he went, or you didn't."

I gripped her hand tighter. I didn't know if Dex heard any of it.

"I'll see you when you come back," she said. "You're going to come back."

We were on the dock where the journey started. They took us to the cutter. They put us into dry suits. Dex's didn't fit. They strapped us into a windowless hold with seats and stretchers and a sink. We were deep inside the water.

"Feeling OK?" one of the Coast Guardsmen asked me. The other man, six feet tall and stooped, his face pink and blotchy, threw up in the sink.

Later, I would describe him to a young man on the island. He would reply, "I know that guy. He's from Iowa."

But, in the belly of the boat, the first Coast Guardsman leaned over me and said, "It'll be bumpy. Hang on."

It was so dark in that boat, and so loud, that we were of the sea, not on it. I followed the seizure—down into the spiraling whirlpool of my body, the warming ocean and its creatures, my childhood with a chronically ill parent, my parenthood as a chronically ill mother, down into the Gulf of Maine and back to the Long Island Sound. I am following it still.

Book One

LONG ISLAND SOUND, 1980

What would an ocean be without a
monster lurking in the dark? It would
be like sleep without dreams.

—WERNER HERZOG

Longneck Clam

(*Mya arenaria*)

In 1980, the summer my father's kidneys began to fail, I was nine years old. At twilight I lay down on the wet, coarse sand of the Long Island Sound's tide flats, grit digging into my knees, face inches from the damp. I pressed my hands alongside the siphon holes of soft-shelled clams, called longnecks or "piss clams" because they drew in seawater, filtered it for food, then shot it back out in a stream. Press the sand and they'd dig deeper. I played them, an orchestra submerged, while inside the sand the clams burrowed away from me, like they couldn't do from a swift rake in the hands of an adult.

"Monkey!" my father called. I looked up; he stood framed by the rows of cottages, barefoot and in dungarees, a bucket in his hand. "Want to dig?" At dawn and dusk, clam diggers filtered out to hunt for longnecks.

"Yes!" I ran up to him, wet sand on my knees and on my dress.

He grinned. "All right then. Find me a good spot."

I ran to the place where I'd been playing the orchestra. "Here!"

He walked up, kneeled, the dampness spreading from his knees up and down the denim. His bony feet curled for balance, toes digging into the sand, arches exposed to the air, vulnerable and naked. With his rake, he combed the ground where the clams were buried, revealing the animals. I sat next to him and dug in the hole he'd made as it filled from the water table below, sorting out clams in the low-visibility murk and dropping them into the bucket.

Out in front of me, the water of the Sound, what F. Scott Fitzgerald called a "great, wet barnyard," stretched shallow and calm in the low tide. Like my father's body, the water of the Sound faced a rising tide of poison. The factories of Bridgeport, Connecticut, were visible in the haze, and I could see a radial tire marooned by the jetty to the west. Behind me was The Periwinkle, Mac's rented cottage on the stretch of summer places we called Dogshit Beach. The Sound only a train ride from New York City, pollution and development had tarnished and crowded the beach where my mother spent her childhood summers. When she was nine, it was a middle-class vacation oceanfront. She could find horseshoe crabs and mussels and snails and fall asleep to the sound of clear blue waves. By the time I was nine, both the ecosystem and my father were slipping away.

When my father finished searching an area, he got up, sandy hand on hip, and scanned the sandbar for another set of breathing holes.

"There!" I pointed.

He picked up his bucket and moved. I ran to the tidal edge to wash the sand off my hands, then ran back to the next hole—hunting, hunting.

His metal bucket was full of fresh water from the outdoor hose at The Periwinkle, to which he added cornmeal and seawater. In the mixture, the longnecks leaked sand. Free from their holes, the clams' shells seemed ready to splinter and peel away from their delicate bodies. When the bucket was full, my father got up.

"Want to freak out your mother?"

She stood in the sand front yard against the blue dusk, smoking a cigarette. She hated the smell of chowder and left the cottage when he cooked clams. She wouldn't eat sea creatures but loved to show me how to find them; she swam every day at the Sound, denying the damaged water and my father's fading health. The baby of her family, she was beautiful, needy, shy, dependent—a translucent jingle shell chiming in the breeze, breakable under strain.

I nodded. "Can we eat the clams this week?"

"We can eat clams this week. No red tide."

"You're not too tired to cook?"

"I'm not too tired to cook." He looked embarrassed.

He didn't say, *My kidneys are fine today*. He wouldn't. PKD hovered around us unnamed. Medical historians believe that PKD has shadowed humanity throughout our existence. Records of the disease begin in 1586 when the Polish king, Stephen Báthory, died on a hunting trip, and his physicians noted that, on examination, his kidneys were bumpy and enormous. Portraits of Báthory show a huge belly, a common result of enlarged kidneys. He and his family had been afflicted by gout, a companion of chronic kidney disease, where uric acid builds up in the blood. The king died at fifty-three. His father had also been in his fifties when he died, while his brothers did not live to see fifty.

It wasn't until 1888 that the first researcher, Félix Lejars, used the term *polycystic kidney*. In 1899, he first recognized that PKD was heritable. It would take nearly another hundred years for anyone to understand the mechanics of that heritability, why for some it is a storm, some a breeze, and how it has become one of the world's most common genetic illnesses. In the long-term, the illness took a predictable course in patients: abdominal swelling, pain, loss of energy, rising blood pressure, fatigue, infection, loss of urine production, and the end stages of acute kidney failure, which could kill in a week or less. Patients succumbed to chills and fever, fatigue, coma-like sleep, and finally death as the toxins in their blood increased to overwhelm the system.

In 1894, at thirty, my great-grandfather—a prosperous pharmacist and married man with three daughters—was well enough to have "gone out on a grand fishing expedition" on Missouri's Gasconade River, according to the *American Druggist and Pharmaceutical Record*. He became ill four years later. No medicine could help him. He could no longer work. At forty, he died. My grandfather was only a year old. The family fell into poverty.

My grandfather joined the Navy at sixteen, sending money home to his mother, trapped by loss in a Saint Louis tenement. In his thirties he fell ill. Kidney disease was suspected, and experimentation recommended. My grandfather underwent invasive procedures to lance cysts, but they only filled up again and grew larger. His doctors used sulfa drugs to treat the fluid as if it was a bacterial infection, which brought only short-term relief.

When my grandfather died (in a hospital room down the hall from an experimental dialysis machine, according to family lore), my father was sixteen. The cycle continued: My father left school, worked to support his mother and siblings, finished his GED at night. His sister Lois, eighteen, gave up a math scholarship at an Ivy League college, got a job in New York City. At twenty-five she'd fall ill. The Caffall Curse was met mostly in silence, in magical thinking, in overwhelm.

Our global marine ecosystem works not dissimilarly from the circulatory system in the human body. Its waters—and the creatures within them—regulate temperature, weather, the food chain. The Long Island Sound, like all estuaries, is a filter, something like a kidney, and also a cradle, a liminal space between the land and the sea. When an estuary is sick from pollution, as the Sound was, it can give rise to life-threatening crises such as red tides.

Red tides are massive marine blooms of algae, so-called because they turn the ocean rust-colored and indicate that the shellfish are not safe to eat. Before colonization, red tides were part of the ecosystem, and Indigenous peoples used them to mark seasons, changing their fishing practices to avoid ill effects. In 1800, the politician Daniel Webster called the healthy Sound "the American Mediterranean"—fish so plentiful you could walk north along the rivers on a highway of oyster shells, eating clams the size of loaves of bread, rivers silvered with spawning alewives. As settlers used and fouled it, writing it off as a waterway for farmers, industry, and ferries, it became less a churning, contained nursery of abundant life and more an instrumental sea. Fitzgerald, looking at it from Long Island, said the Sound was "the most domesticated body of saltwater in the Western Hemisphere."

When the water is saturated with an overabundance of nitrogen sloughed off farm fields in the rain and leached sewage from overburdened waste facilities, it is called over-enriched, as if capitalism's goals have spilled over into the sea.

A teaspoon of seawater can contain millions of individual plankton and hundreds of plankton species: phytoplankton, tiny water-going plants, and zooplankton, tiny animals that feed on them. Together, they are the drivers of the food web, building blocks of the world's oceans. Phytoplankton produce about half of our planetary oxygen, as much per year as land plants, and sequester a significant portion of our carbon. A red tide blooms the ruddy dinoflagellate microalgae *Alexandrium fundyense*, a type of plankton. This is sometimes called a harmful algal bloom, or HAB. These plankton house potent neurotoxins and saxitoxins, exposure to which in humans can produce paralytic shellfish poisoning (PSP), which can lead to paralysis, to death.

Plankton derives from the Greek word *planktos,* meaning "made to wander or drift." Any ocean creatures that cannot push back against currents are considered plankton, from blue-green algae to jellyfish to the one-ton ocean sunfish. Phytoplankton like *A. fundyense* float, consuming sunlight, using photosynthesis. These building blocks of earthly life, plant and animal, have near-crystalline variety, form, function. They branch like vines, blossom, are armored, are finned like orchids or faceless fishes. Diatoms, green algae, coccolithophores, dinoflagellates, ostracods, copepods, arrow worms, by-the-wind sailors, and radiolarians bloom as translucent oddities—a parade of tentacles, pointed stars, clawed aliens.

Red tides can feel like a curse. In the *New York Times* in August 1975, Bernard Feder published an article looking at the growing prevalence of red tides in the United States. "When the Red Tide Strikes. . ." describes a disastrous Atlantic red tide: "In September 1971, after hundreds of dead and dying seagulls and ducks were seen off the coast of Gloucester, Mass., scientists ascertained that the birds were victims of PSP from a dinoflagellate bloom." The catastrophe had spread, like localized smaller red tides that first closed Yankee fisheries in 1970. But this tide "included the closing of shellfish industries in New Hampshire and Maine. New England was declared a disaster area so that clam diggers and shellfish handlers could collect unemployment insurance."

Feder's epigraph was: "*And Moses . . . struck the water of the Nile . . . and all the water was changed into blood. The fish died and the river stank, and the Egyptians could not drink water from the Nile*" (*Exodus 7:20.21*).

Red tides are biblical, visible, ongoing, season to season. Genetic legacy extended our story not only forward and back in a straight line of inheritance, but as a spreading tide.

In a study of a single multigenerational Vermont family with PKD during 1973 and 1974, researchers from the psychiatry department of the University of Vermont found that a well of denial ran underneath their experience. In the abstract of their paper for *The International Journal of Psychiatry in Medicine* titled "Individual and Family Coping in Polycystic Kidney Disease," Drs. Delia Mary Manjoney and F. Patrick McKegney found that the denial changed how individuals acted in the face of the crisis. They subtitled their paper, "A Harvest of Denial." The authors wrote of one subject that she "had maintained a strong denial of the possibility of having the same kind of kidney disease which had killed her mother. She 'didn't understand' her father's plea that she not have more children."

A. fundyense respond to over-enrichment by breeding and depositing seeds that will bloom in red tides the following summer. A filter feeder like the longneck clam is especially likely to consume *A. fundyense*. Measuring these seeds can enable scientists to predict what will come in the future—a healthy year or a year in which the clams can kill.

The seeds of *A. fundyense* are called cysts.

As the Housatonic River delivered nitrogen and toxins to the estuary from factories and farms, it delivered my family to the Sound. We flowed with it, back and forth from our house in Great Barrington, Massachusetts, to Dogshit Beach. My mother's mother owned the building we lived in, over an Exxon station; my father made reproduction Shaker furniture in the shop next door, stored lumber in the barn. The Exxon station had old-style pumps, a sign facing Route 23 with the trademarked tiger, a store, and the two apartments—one above the store, one behind it—all abutting a long meadow at the foot of East Mountain. My grandmother's apartment was renovated. Our apartment was cramped, narrow, and dark, with indoor-outdoor carpeting and linoleum, paneling, and cast-off furniture pieces built by my father's apprentices. He'd promised to fix it up. He never did. "The shoemaker's children go barefoot," my mother said.

Mac's money rumbled at the edge of our poverty. She was tall, broad-shouldered, a beak nose, a serious haircut, muscled from years as a swimming instructor. Raised in the Midwest, swimming in Lake Michigan, she came east for graduate school, studied the metabolism of posture, wrote about it in a PhD thesis used by NASA for astronaut training, and married Doug, the philosophy PhD student who became my grandfather. Doug was a rail-thin academic, in tiny round glasses and slicked-back hair with little interest in sports. Mac was a tanned athlete. With her marriage came the only name I heard anyone use, not Harriet, but Mac for Doug's last name, McCormick. She taught college sports physiology, and he became an editor at Prentice Hall—a line of socialists behind her, a line of clergy and scholars behind him. My mother couldn't keep up with her father's lectures on literature, didn't read quickly, didn't like to write. Her dyslexia wouldn't be diagnosed until she was an adult. Her sister was the smart one. Her scientist mother couldn't see the scientist in her child. The daughters split, my mother to Mac, her sister, June, to Doug.

When Doug died young of throat cancer, Mac took a year and traveled the world. Our attic held carousels of color-saturated slides, pictures of her, tall and in orange polyester pants, standing in front of pyramids, volcanoes, oceans. She read stacks of science books, Rachel Carson, Richard Leakey, and glued two slices of nautilus shell on the door to her apartment in our house in Great Barrington, so the light came through the spirals. She sailed every summer up to the Gulf of Maine. She knew better, always.

When she was away and we were responsible for the bills, our electricity was turned off. When we didn't have money for food, we ate government cheese. I learned to shoot a rifle in the woods with my father, learned to track deer, learned to fish, learned to eat what he could hunt when we were hungry—a hybrid of conservation ethics and poverty, creating murky lines of morality.

I didn't make friends. I didn't understand the rules other children had. I couldn't lie. I only thought about animals. I spoke fast, used vocabulary adults favored, read adult books, wrote adventure stories. Numbers and musical notes reversed themselves on the page, but I could quote book passages from memory and play flute by ear. I couldn't get right the games or sports or dances or gossip of the playground, couldn't read facial expressions or body language. Outside of every joke, every group, I spent hours watching frog spawn, looking for crickets in the meadow, setting out plates of nuts for the chipmunk in our woodpile, reading *Ranger Rick* and *National Geographic* magazines in the barn, cutting out pictures of orcas, humpbacks, dolphins, seals. I made maps of my forest as if it were a kingdom, created a world just to the left of the one where all the other children lived.

My ability to talk to adults made it easier to take me to dinner parties when there was no money for babysitters. My mind was like my mother's, like Mac's. My mother took dictation for my stories before I could write, took me to the library, to museums. Mac read to me about ecosystems and star systems. My mother told me about the history of the hills where we lived. My father told me that fitting in was overrated.

I kept quiet and taught myself to be tough, more my father's daughter than my mother's. She had been the sick child in her family, born premature when, as she told it, her sister leapt onto Mac's pregnant belly one afternoon while they sunbathed at Dogshit Beach. Another summer she was rushed from their cottage to a hospital for a tonsillectomy where she was isolated from her mother for two days; she came back angry and resentful, Mac guilty and doting. She stuck close to Mac, quiet, like her, curious, like her, alive only in water, like her. My mother didn't know how to take care of anyone else, and my father wouldn't tell her how to take care of him. Through silence, the Caffalls survived in a world without legal protections for the chronically ill, where admitting their incurable disease could mean loss of hope, work, and anything the new generation built on the ruins of the old. My mother embraced the attention that came with illness; my father resisted it.

I knew about the Caffall Curse, but only sideways. At holiday dinner tables, my father's siblings spoke in code, whispered in hallways outside my door when I was supposed to be sleeping. When I asked, they told me it was a disease some of our family had that made bubbles in the kidneys, filled with water. They spoke as if it wouldn't touch me, a flooded land they inhabited that I could see from the dry shore. It was as inexplicable to me as the causes of red tide.

One afternoon in that summer of 1980, I found my father asleep in the gas station, his torso draped over an angled drafting table. In the woodshop north of the main house, his apprentice worked on a Shaker cabinet; in the storefront—with the cash register, case of snacks, and shelves of antiques for sale—my father worked on designs while he waited for the afternoon rush of cars heading from town into the hills. "Are you all right?" I asked, my hand on his shoulder. He sat up, revealing a drawing of a grand-father clock. His pencil rolled onto the floor.

"Oh, yeah, of course," he said.

"You fell asleep." I picked up the pencil and tucked it into the pocket of his denim shirt next to the notebook where he kept lists of the books he wanted to request from the town library.

"I did." He rubbed a hand over his eyes, blinked at the empty storefront. "No customers?"

"Just me. I wanted a snack." I walked to the cash register, took a candy bar from behind the counter.

The day before, he'd asked me to play cribbage while, out the window, the afternoon sun caught the face of Monument Mountain. "Do you know that the rainbow trout we caught last weekend wasn't even from here?" he asked, animated by just the thought of being on the river. "Those fish are supposed to be in western streams, not the Konkapot." He volunteered at Trout Unlimited, and he'd started going to lectures, taking me with him.

"Does it think it's a trout without limits?" I asked.

"There's my smartass!" he crowed, a grin lighting up his face. "Now see if you can skunk me."

But that afternoon he didn't ask me to play. "Just tired?" I asked.

"Nothing to worry about, Monkey." He put his hand to his back and massaged the side of his body, wincing. He had circles under his eyes. "I'll be OK." From the windowsill next to his drafting table, he fished out an aspirin bottle. I remembered him telling me once that his father worked until the week before he died.

That night, Mac, my mother, and I watched *The Undersea World of Jacques Cousteau* on PBS. It was dark outside. My mother and I sprawled together on Mac's scratchy mid-century couch, a fire in the woodstove, the Milky Way spiraling around me in the imaginary dark of my closed eyes. "Long before there was a sprig of grass on Earth, there was life in the water," Cousteau intoned. I pretended that I could stay in that warm room all night, wake to the view of East Mountain from her sliding glass doors, peer through the nautilus shells she'd glued to the glass, looking at the green lilac trees.

"You could go back to school," I heard Mac say, her voice floating in swirls of gas clouds and nebulae somewhere distant. My mother was a model before my parents left New York City when I was born. In the Berkshires, she tried many things— writing books on quilting, teaching needlepoint, cooking. She was endlessly curious, quick to learn new skills, with a brilliance for landscapes and folk art. But she bathed in anxiety, afraid of heights, horses, crowds, talking on the telephone, losing my father, being poor. She was a slow reader. The other systems that made her mind unique were dismissed. She was so beautiful that her social panic was written off as quirk. She helped pay our bills by cooking in hotel kitchens and lived for the water. "You know," Mac went on, "finish your degree. You can't cook at inns forever."

I could feel my mother's body tense. "What would I study?"

"What do you like?"

"The ocean, I guess." She was quiet. "Where would I go?" Her voice sounded like packing a bag.

"Well, UMass isn't that far away."

"Two hours."

"You could make it work."

"What about the business?"

"He won't be able to work much longer."

I opened my eyes. No one had ever said that near me, and I lined up facts like points in constellations—my father asleep on the drafting table, his sore back, his fatigue after fishing, hunting, walking the forests.

"I know," she said, like standing in the crook of an open car door, turning to wave goodbye. She was silent for a long time. Then: "I wonder if they have oceanography."

Later, in our own living room, while my father slept and my mother and I were on our ratty couch, too dark to see anything but the headlights of passing cars, she asked me: "Would you be OK if I went back to school?" as if it just occurred to her. "Because Daddy is getting sick." She almost whispered.

"With what Aunt Lois has? With what his dad had?"

"Yes. His family all has it."

"Uncle Brian?"

"Well, we don't know yet. He's ten years younger. But Patrick does."

"What about me?"

"We don't know about you either. But you could."

I froze. "OK." I was starting to burrow under sand, under water.

"You have a fifty-fifty chance. So, maybe not. Maybe it won't happen at all."

Fifty-fifty. I wasn't sure I knew what that meant. She had said it and now it was mine; I would have to hold it. I could imagine how she'd look at me, the way she looked at my father. Already, she was looking at me like that in the dark.

In my father's lumber barn, among the planks of curing tiger maple that would become Shaker tables, I made a corner into a *Ranger Rick* club, with my pictures of whales and seals and dolphins from magazines taped to the walls, bare bulb suspended over a dirt floor, a pile of lumber stacked and covered by a quilted moving blanket. At nine, I knew animals were in danger. At the Trout Unlimited lectures and river cleanup days, they talked about acid rain. The back of our car was littered with his pamphlets on pollution. My mother read Rachel Carson's books too. Mac donated to Greenpeace. They rarely spoke about what was collapsing or why, only told me that it was.

I didn't understand that the chemical compounds of acid rain were contributing to the increase in red tides, dissolved nitrogen entering the atmosphere at increasing rates as the water cycle drew it up from fertilized fields, then sent it back as rainwater. Whales and seals relied on the food sources impacted by those forces. The illness of the streams around me seemed like something we could correct on weekends of river cleanups, something that could be healed if we just bent toward the work. The terrible danger to the ocean, the endless chain of toxins reaching from the tiny rivulet that ran down East Mountain in the springtime to the Sound— that was missing from my consciousness.

When we caught the rainbow trout, we were on the Konkapot River, my father's favorite, a stream that fed the Sound. We drove into the hills and walked up the pine-carpeted stones along Umpachene Falls, ate sandwiches while our feet soaked in a shallow trough of moving water, then drove to a dirt road where he parked on the side, pulled waders and a fly rod from the trunk. I took my book and walked with him through the brush to a pool. "No one knows this spot," he said. "I found it when I was working on that house out here."

I trailed him, watching out for brambles, picking fat blackcap raspberries and putting them into my mouth, sun-warmed and sweet. "Are there fish?" I asked.

"Rainbows," he said, grinning back at me, "maybe brookies." The stream was rocky. It was hot, and I took my shoes off to wade into the water. "Quiet," he whispered, "don't want to spook them." He put on his waders and prepped the fly rod, choosing a fly he'd made himself; a new hobby that required no standing, no lifting two-by-fours, a hobby for exhausted days. But on that good day, he didn't need rest: He stood tall as the sun lowered and mayflies hovered over the water and the trout came for them.

A week later, my parents loaded up the car and the three of us followed my grandmother to the beach. My father seemed tired and distracted. "Don't kick the seat," he said as he drove south along the Housatonic, "my back hurts."

I understood what was supposed to come next for my father, when his kidneys no longer filtered poison from his blood: dialysis, the most common intervention in the United States in 1980.

My Aunt Lois had been on dialysis for my entire life. She'd shown symptoms of PKD early, and kidney failure sent her into early menopause. She started dialysis in a hospital in New York City, then dialyzed at home. Mostly, she made it look easy. While she had her blood cleaned, she read books and sewed her own clothes—loose, to hide the bulge of her enormous kidneys. She worked in an advertising agency on Madison Avenue that had the accounts of the Metropolitan Opera, the New York City Ballet, the Bronx Zoo. When I visited, she produced tickets and trips, then disappeared into her treatment, returning tired. She taught me how to cook, but in conversation with her brothers, she'd list in detail the restrictions on what she could eat and drink, how to control protein, potassium, phosphorus—chemicals that would build up in the blood, fluids that would choke the system. Dialysis saved her life. Dialysis became her life.

In dialysis, the patient's over-enriched blood is run into a series of permeable membranes, bathed in a solution that splits the chemicals from the blood. Magnesium, sodium, potassium, sulfate, calcium, chloride, and phosphorus are removed until the blood is cleaned and returned. By the 1960s, technological and plastics innovations enabled doctors to improve the procedure and make it more widely available. There was only one problem: So many people were in chronic kidney failure, yet so few could afford the treatments, and without it their life expectancy was under two weeks. Who was allowed treatment, and who died without it, became a question of scale and will, of the worth of a vulnerable life, of the rights of the chronically ill.

Between the 1960s and the year I was born, the process of getting dialysis was a nightmare. Panels at individual hospitals were created to assess those in need and choose who could have the treatment. Patients were evaluated by income, marital status, number of children they were supporting, physical strength, earning potential, age, weight, mental abilities. These Admissions and Policy Committees—sometimes called God Panels—were made up of ordinary community members, clergy, labor leaders, businessmen, housewives. "Agonizing practical decisions must be made," Shana Alexander wrote in a 1962 *Life* magazine exposé on the panels in one Seattle hospital. "Someone must choose which one patient out of 50 shall be permitted to hook up to Seattle's life-giving machines and which shall be denied." The article was so incendiary that there was a subsequent NBC documentary and a mass outcry against rationing lifesaving care. At the time, it cost $15,000 to $20,000 a year to save the life of a patient with end-stage renal disease, or ESRD.

The country was wrestling with what it meant to have a burgeoning ability to save lives, and no political will to protect them with universal health care or socialized medicine. In 1967, a federal committee recommended that the United States amend the Social Security Act to protect patients with ESRD from the costs of dialysis. During the congressional hearings, representatives heard testimony from Shep Glazer, a dialysis patient from New

York. "Gentlemen, what should I do? End it all and die?" Glazer asked. "Sell my house for which I worked so hard, and go on welfare? Should I go into the hospital under my hospitalization policy, then I cannot work? Please tell me. If your kidneys failed tomorrow, wouldn't you want the opportunity to live? Wouldn't you want to see your children grow up?" During his testimony, Glazer was briefly dialyzed on the floor of Congress. President Richard Nixon signed the amendment into law in October 1972, extending Medicare coverage to people with ESRD through Section 2991 of the Social Security Amendments of 1972. Some consider it the closest the United States has come to socialized medicine: comprehensive health care for a single organ.

At that same moment, Americans were wrestling with new ideas of interconnectedness and interdependency, stretching to think about how our choices to protect public health overlapped with our responsibility for environmental health. While the debate about the worth and dignity of vulnerable life raged in biomedical ethics and political circles, a similar debate was underway in environmental ones.

In 1969, *Time* magazine published a report on riverine pollution in Cleveland's Cuyahoga River, which had recently and famously caught fire. "Chocolate-brown, oily, bubbling with subsurface gases, it oozes rather than flows," the magazine described the river. "It is also—literally—a fire hazard." With its roundup of the grotesque ("animal grease balls as big as oranges"), the article was nearly as incendiary as the river pollution itself, and it changed the American conversation about clean water. The Cuyahoga River fire became linked in the public consciousness with the growing environmental movement, the revelations of Carson's *Silent Spring*, and other environmental disasters that had littered the decade—four hundred New Yorkers dead from smog in 1963, a four-hundred-square-mile oil spill in 1969 on the California coast.

By January 1970, as public opinion in favor of action was increasing, Richard Nixon said in his State of the Union address, "The great question of the seventies is, shall we surrender to our surroundings, or shall we make our peace with nature and begin to make reparations for the damage we have done to our air, to our land and to our water?"

It was in this climate, in 1972, that the Clean Water Act passed. In his remarks before the Senate vote, the bill's sponsor, Edmund Muskie of Maine, painted a dire picture. "[T]oday, the rivers of this country serve as little more than sewers to the seas. Wastes from cities and towns, from farms and forests, from mining and manufacturing, foul the streams, poison the estuaries, threaten the life of the ocean depths. The danger to health . . . can be anywhere."

With the passage of the Clean Water Act, my family's beloved Sound had a chance at restoration. By 1980, the Sound should have been seeing improvements from that regulation, but it was about to enter a period of terrible danger—red tides, hypoxia, anoxia, murkier waters, swimmers with rashes—that would culminate, though not end, with a mass hypoxic event in 1987.

An estuary is a singular ecosystem, as productive of life as a coral reef, a rainforest. The abundance of creatures stems from its richness in diatoms, single-celled planktonic algae, microscopic plants born in those liminal spaces between sea and land. As Tom Andersen points out in his book *This Fine Piece of Water: An Environmental History of Long Island Sound*, estuaries are intense, alive places, "producing twenty times more organic material" than oceans. The name itself is a rolling thing, from the Latin *estuare*, meaning, he wrote, "to heave, to boil, to surge, to be in commotion." They teem with creatures in a food chain depending on richness: on winter ice that shears spartina grass off from the salt marshes, on those grasses floating into the Sound and decomposing, adding food for diatoms who feed plankton, who feed larval fish and lobsters, who feed adult fish, who feed seals and whales. Estuaries can give a snapshot of oceanic health, help us predict what's coming, but the subtle changes of a place or a body can go unnoticed, or can bloom and fade as disaster approaches.

A PKD kidney works not dissimilarly. The cysts themselves do not release toxins. A few of them are survivable. Many adults die of natural causes with cysts in their organs. Cysts themselves are uncomfortable, disordered, but not deadly. But as the disease progresses, they increase in number in the organs and would be visible if the kidneys were removed from the body. If you dissect a PKD kidney, the cysts reveal themselves all through the organ, on every nephron, in tissue that is otherwise ordered, until they roil the body, enriching it with everything the patient eats, everything it cannot remove—all the things that made the body healthy are there, but in overabundance.

The clams in our bucket that day on the beach weren't damaged by red tide, though they may have been filtering sewage and heavy metals and mercury and dioxin and PCBs and PFAS and DDT. I sat on the beach while my father sorted through the clams, watching the sun set, making drippy castles with the wet sand at my feet as I thought about chowder and the time we'd have in the kitchen, just the two of us, and how good it was when he had the energy I remembered.

I remembered him the year before, when he was still strong. I could still feel him lifting me onto his shoulders on that beach at sunset, walking to the water to point out sailboats he'd want if he could afford them, the song he'd taught me, "Simple Gifts," the one he described the Shakers singing in ecstasy, in a trance while they danced in circles:

> *When true simplicity is gain'd,*
> *To bow and to bend we will not be asham'd,*
> *To turn, turn will be our delight,*
> *Till by turning, turning we come 'round right.*

He made it look like a practice, turning and bending, ignoring pain and the future, his silence like a reed in the water. You lived with risk and fear, moved and swayed with it, courted it like a dancer in a Shaker trance, like a clam disappearing below the sand. You hid, you emerged, you waited out danger and suffocation for a moment of transcendence. He wanted me to understand that without ever saying a word.

I watched my father walking up the beach with the bucket. "Ready?" he yelled, smiling. He was shimmering somehow, his freckles blending together to imitate a tan, his glasses glinting, lit from the golden hour. He held the bucket in his right hand, the delineated muscles of his forearm powerful. He was familiar. I was like him.

I nodded. "Ready!" But I wasn't ready at all.

Long-Armed Hermit Crab

(*Pagurus longicarpus*)

I floated in the Sound; Mac held her arm under my back. "If you can breathe, breathe," she said. "You have to become part of the water, or you'll panic."

Over me washed the sunlight. The smell of Coppertone mixed with salt. There was an oil slick around my body where the sunscreen washed off. There was a seagull overhead when I opened my eyes. There was a closeness of Mac's body, the brown of her arms. Her thin hands were like my hands, like my mother's hands, like my Aunt June's hands.

"I don't know how to float."

"Look, I'll take my hands away, and you'll just do it."

I felt her take her hand from under my back. My bottom began to sag toward the sand, and I wanted to reach down with my toes. "I feel heavy."

"You are heavy. But the water is stronger than you are, so you have to trust it."

"What if I can't trust it."

"You can. I promise you that you can."

I felt downward with my foot. If I could just touch the sand.

"Don't cheat. Have faith."

At breakfast that morning, my mother asked me, "Do you want to go for a swim?" Dishes were still on the table, the *Times* in pieces on the floor, my feet covered in sand, and the soles making sandpaper noises when I bounced them under the table. The Periwinkle hovered over the sand—sand for a yard, sand on every floor, sand in our teeth, sand in our hair—three tiny bedrooms and an outdoor shower with a screened porch with a daybed and wooden picnic table that looked out at the Sound. Something in the way my mother asked made me nervous, the insistence that I swim.

"Can we go for a walk to the reef?" I asked.

"Maybe before dinner."

I was afraid of the deep water at the Sound. In ponds and rivers at home, in creeks and lakes, I could see the bottom and the other side, the depths of the water finite, the edges contained. In the ocean, if I kept walking there would only be more water, until I was floating at the top of a world that stretched below me, utterly unknowable, powerful, vast.

"Could we get shells?"

"Don't you want to swim?" Mac asked. I shrugged. "You know how to swim." I shrugged again. "Didn't you get her any lessons?" Mac asked, turning to my mother where she'd stood up to clear the dishes. Over nine summers, Mac had walked with me on the wooden walkway that stretched from the road to The Periwinkle to the water, passing cottages that went from small to large like a line of periwinkle shells. She swung me over the waves when I was small enough to carry, making my toes graze them. She'd seen me dog-paddle and float on the inner tubes of semitruck tires. She never noticed I could not swim. She didn't know I was terrified of the depths of the sea, that it felt like I was falling off a building when I knew I couldn't touch the sand below me.

My mother ducked her head. Part of the *Times* was on the table at Mac's elbow; my father held another section, silently filling in the crossword. He knew the trap of answering Mac's question. "We couldn't afford it," my mother said. Her defiant tone was the one I loved, the one that protected my strangeness from the world. It was the way she spoke to anyone who didn't understand me. She and I were disconnected often, looped and failing, but if anyone saw me wrong sized, she came alive. She explained me to the town librarian who told me I was too young to check out books from the adult section: "She's been reading since she was three. She could read rings around anyone here. She'll check out what she likes."

My father looked up from the section of the *Times* he was reading. Mac put down her fork, the last eggs like crumbs on her plate. "No lessons at all?" she asked.

"No."

"This afternoon," Mac said, "before low tide. We'll start then."

My mother got taller, her shoulders went back. "OK. But then later, sweetie"—she turned to me—"before dinner, we'll go to the reef." Her voice the protective one she'd used with the teacher at the private school I'd gone to for only a year, the one my father traded handmade furniture for my tuition. They had rules about how old you could be before you were allowed to read, how still you had to lie at naptime, shaking hands with the teachers when you left for the day. I could already read. I didn't nap. I was told to lie about both. I couldn't manage any deception or the sleep and broke a plastic necklace I'd been using as a fidget while I stared at the ceiling one naptime, sending tiny beads all over the floor, waking all the students. The teacher reprimanded me in front of the whole class, and I refused to shake her hand at pickup.

When my mother arrived, she hid me behind her skirt. "She's not going to shake your hand if she doesn't want to," she said, "you must have done something terrible."

Mac met me at the sand yard and didn't say a word. I wanted my body to do what I asked it, to swim with the long strokes I saw in my mother's body, my aunt's, my grandmother's. They lived in their bodies, athletes, swimmers, dancers. I lived outside of mine, begging it to comply, rejected every time. I could feel the old sunburn peeling on my shoulders as we walked into the water until we stood up to our waists in the Sound.

"Stand up straight." She looked at me over heavy, black eyeglass frames, her swimsuit the kind swimmers wore in competitions. I straightened. She took a strand of hair out of my braid. She gently extended it into the sky. Still attached to my scalp, it pulled softly on my skin. "Pretend there is a string from the top of your head down to your heels," she said. "It runs in a perfect line, and lines up your spine like a set of blocks, so you can't slouch." I adjusted again. She looked at my body refracted in the water. "Now you look like a duck, with your butt sticking out like that." I kept my feet rooted to the bottom as she spoke, my toes gripping the smooth sand. I knew the sand; I knew the flats. "OK," she said, "show me how you swim out."

I dog-paddled away. Below me I imagined hermit crabs walking. I tried to move like she moved in the water. I paused to reach below me with my feet to feel the sand. The sea reached my ribs at the first check, then my shoulders, then my stretching neck, then my chin pointed to the sky for air, my toes searching. The Sound stretched out in front of me. At the last moment before the shore dropped to the ocean bottom, my feet touched nothing, vertigo took hold, floating became falling, drowning, an ancient memory of storms and ships. Anything could be below me, animals, plankton, seaweed, stones.

"Slow down!" Mac called from the shallows.

I couldn't.

On land the ocean was a comforting enormity, the surface a smooth glass plate, a roiling tumble, a dance of low peaks; I longed to disappear into it, an animal, a swimmer. Inside the ocean, it was something else, a bottomless void, a loss of control, a swallowing edge of mystery. I loved what was hidden under the surface. I feared what was hidden under that surface.

"Come back here," Mac yelled, alarm in her voice. "You can't swim if you're panicking." I turned away from the view of the open water and saw Mac walking toward me with the heavy gait of water pulling at her legs. "What happened?" she asked when she reached me, tall enough to support my legs in her arms; the water covered her shoulders.

"I don't know."

"Do you know any strokes?"

"No." I didn't tell her how frightened I was.

"OK." She softened. "We'll start at the beginning." She picked me up in her arms, turned me onto my back, my tense legs at an angle until she stretched my body out and walked back with me, her hands under the small of my back, the water shallower and shallower until I rested on its surface at the level of her waist. "Close your eyes. This is how you float." She let me hover there, sun on my face, the strong bones of her hand pressing me up like a solid wooden chair, until I relaxed into her, peace warming my limbs, shame liquid in my throat. "Just float for now," she said. And I tried to let go of anything but the sound of birds, the idea of crabs on the sand beneath. Finally, the receding tide sent us back to the shore. "We'll try again tomorrow," she said as she let go of my legs and I found the flats, my knees wobbling with released fear.

We stood facing each other. She looked me over, and I recognized the way my grandmother searched my body for signs of failure, an unconscious habit of my mother's family, an appraisal—she was a water goddess I'd disappointed.

"When are Nolan and Quincy coming? Aunt June is coming, right?" My cousins were boys, one older and one younger. When we were all in the cottage, it was messy and loud, one cousin slightly mean, one deeply kind. Mac adored them, swam with them. I coaxed them to play pretend in costumes we improvised out of terry cloth beach shirts and floppy hats, holding water pistols, detectives on a tough case. I read books in the cool sand under The Periwinkle when they came, weird sister, in the crawl space with the cinderblock foundation pillars around me and the smell of the privet hedge, the loud bluster of them in the house above comforting.

"They'll be here from Storrs on Friday."

They didn't have it, I realized. On my mother's side, there was no illness. On my mother's side were strange people, talkative, busy, focused, blunt. On my mother's side were PhDs and research trips, professors. On my mother's side was health. On my mother's side was also a unique kind of mind.

"You can learn this."

"OK."

"There isn't anything to be afraid of."

We walked back to the cottage. I sat on the daybed, looking at the toys in the sand yard, exhausted. I could hear Mac walk to the back of the house, turn on the outdoor shower, wash off, and come back to the porch.

"You'll get it," she said from the doorway. "It isn't easy for some people."

"It's easy for you."

She walked up to me and put a hand on my arm. "You aren't me. Go play." She kissed me on the top of my head.

I opened the screen door, heard it slam shut behind me, but didn't look back at Mac. I grabbed a bucket from the yard and walked alone the same way we'd gone out to the lesson. The flats were emerging as the tide left the shore naked, and I could see the long sandy tide pools there. Somewhere in them were long-armed hermit crabs.

Before we left for Dogshit Beach, fogs of argument rolled from my parents' bedroom in Great Barrington. "Why can't we make love anymore, Charlie?"

"Because, Nancy, I am tired."

"I can't do this, Charlie, I can't." My father's body, as always, the problem.

I wanted to not be a body at all, to be invisible. In the woods I couldn't hear them. Lying in my bed, I pictured the deer I could see in the meadow out the back window of the spare room, took my blanket and book and went to the bed there, under that window, to lie awake farther from their voices, their fury and disappointment, and watch for the deer until dawn. I wanted to be able to change into one, to give over the ache in my chest, the way I was supposed to be close to my mother, like her and like my grandmother, pushing back against the man who thwarted what they wanted. But I was like him instead, like the deer he was teaching me to hunt. My eyes wide open, I wanted nothing more than to run. I wasn't like my mother. She was beautiful. I was not. She assumed she'd have everything she wanted. I taught myself to want only the forest and the ocean. I might be sick one day, a spell broken only by finding out you are really a deer.

I tried to breathe. I listened to her go into the bathroom, then into their bedroom, the hollow-core door closing dull and empty, no boundary and no protection. The moon was over the hill, and I saw shining eyes in the meadow in the dark, raccoons or opossums, moving low through the high grass, the barn in shadow. I fell asleep trying to guess what the animals were. In a nightmare I was chased by a witch while I ran up and down stairs like those in the barns of the Hancock Shaker Village where my father went sometimes for work. The stairs reversed and twisted, falling down and endless, while I ran.

In the morning, cool light filtered down from the hills. My body was somewhere below me, an unruly thing that could hold betrayal, as far away as sand below the surface of deep ocean. I left the bed, walked into the hall, the sunlight on my feet. I walked into the living room and sat next to the black-and-white set, turned it on and dialed the volume so low I could barely hear it a foot away, changing the channels, looking for Cousteau and whales and reefs, people in dive suits, breathing air, no ocean bottom in sight, but safe, contained in a private bubble, peering through a pane of glass.

Instead, on channel 11, the 1934 film *Treasure Island*—it was nighttime, a fleeing pirate saw Captain Long John Silver coming into the Admiral Benbow Inn disguised as a beggar, the pirate ran from him, fell dead, fell down the stairs, the boy, Jim, found a slip of paper in his hand. It read, "You have until ten tonight," below a large black spot. My heart raced as Jim read the note, and as he rushed to bolt the door. Panic flooded to the end of my fingers. The black spot was a whirlpool—empty and echoing. There was too much below me, a depth, a blackness. I could fall into the paper in Jim's hand, into a threat and a promise. I stood up and turned the TV off, breathing hard as if I had been marked and not the page. I was sweating.

The Housatonic, the river that ran only a quarter of a mile past our house, is one of three rivers that feed the Long Island Sound

90 percent of its freshwater. The Sound's watershed is enormous, with 16,570 square miles of drainage feeding its 1,220-square-mile area. Some of the most PCB-dense areas of the Sound's watershed can be found at the mouth of the Housatonic, a legacy of industrial use begun with the region's first water-powered grist mill erected in 1739 in Great Barrington, Massachusetts, and stretching through leather, wool, paper, and copper mills, as well as agricultural runoff.

In one area alone—the PCB-contaminated site of the General Electric plant in Pittsfield, Massachusetts—the riverine pollution is staggering, especially at Woods Pond just below the plant. "Currently, more than 50% of the PCBs that enter Woods Pond go over the dam and continue downstream, even into Connecticut. A distance of approximately 140 miles," says the United States Environmental Protection Agency website. "PCBs are present in large quantities in river sediment and floodplain soil; estimates range from between 100,000 to nearly 600,000 pounds of PCBs.... Without cleanup, it would take decades if not hundreds of years, before PCB concentrations in fish would decrease to a level that would permit unlimited consumption."

The plant that sent PCBs into the river from 1935 to 1977 was designated a Superfund site in 1997 and has undergone extensive remediation, but retains high levels.

The Connecticut section of the river has lingering mercury, a result of the hat-manufacturing industry that flourished there in the nineteenth and twentieth centuries. Remediation to remove these poisons would involve the removal of contaminated sediment from the river and the Sound, and the storage of it in special landfill facilities proposed for the Berkshire Hills. How that would be conducted is still the subject of local lawsuits.

By 1980, the Sound was only a ghost of what it had been at first settler contact—cloudy water, hypoxia, red tides, brown tides, missing animals, full of bacteria, radiating loss, courting suffocation—but kept its deep color in the morning, waves on storm days, smooth, pond-stillness under July's hot, white sky. We brought our darkness up and down the course of the Housatonic River, along the route of collapse.

I walked downstairs to Mac. She was packing for Dogshit Beach.

"Can I go with you?"

"You're coming at the end of the week."

"Why aren't we going the same day?"

"Your father has to work this year."

"I want to come now."

"You're moping. Go play."

I opened the sliding door, nautilus shells from an ocean far away winking in the sunlight. I walked to the barn, turned on the light over the section where I'd set up my *Ranger Rick* Club. Since starting it, I tried to invite children from school—the immigrant's daughter, the farmer's daughter, the quiet boy.

"I think we should work on acid rain," I told Yolanda at our first meeting.

"Is acid really going to fall from the sky?" she asked.

"It already is!"

"It is?"

"It's already in the lakes, and the fish are dying."

"Like Lake Mansfield?"

I nodded.

"I thought we were worried about whales," she said.

I said, "We're worried about all of it."

They stopped coming. No one felt it like I did. I was the animals. Something stalked their survival. Something stalked mine.

Sitting on a dusty board was a collection of shells from the Sound. On top of the pile, the shell of a northern moon snail, huge and gray. I picked it up and held it in my hand. Like the ladyslipper or the razor clam or the jingle shell, moon snail shells were everywhere on the Sound. We had plates of shells in the house, in the cottage on Dogshit Beach.

A moon snail is a monster. Their shells are round and plump as miniature baseballs, or rather, like gray-tan versions of pink Spaulding balls, as if you could throw them and they would bounce back to be caught in your palm. Even the whorls of their shell tops are smooth, like nipples on a breast laid flat on the chest of a sleeping woman. The openings for the animal are wide enough to fit the pad of your thumb. Moon snails thrive in the Sound, look harmless. But they are slow-motion threats, skillful predators that only seem flaccid. They are muscular—extravagantly bulky bodies covered with extravagantly bulky armor. They burrow into the silty mud of the tidal flats after clams. When they have one, they attach to its shell. Then the moon snail bores a single clean hole to penetrate and destroy. The shells of the dead clams litter the shore as well, each marked with a perfect, circular hole, a black spot.

I looked at the shell, then looked out through the open barn doors. I could see Mac loading her car. Inside the wide interior of the barn was hollow and empty. I couldn't protect anyone—my father, the animals. She'd drive to the sea, and I would be alone with it all.

In "Individual and Family Coping with Polycystic Kidney Disease: The Harvest of Denial," authors Manjoney and McKegney write, "Denial in the spouse of an affected individual hindered discussion about P[C]KD, and substantially reduced the emotional support of the affected person. In effect, denial by the spouse isolated the affected P[C]KD individual, was interpreted as rejection by the affected person, and contributed to significant depressions."

That night, my father fell asleep in his rocking chair, the cat on his growing belly, my mother at the table, poring over ledgers where she kept the books for the furniture business, dinner dishes uncleared. I put myself to bed. That night, I dreamed of the black spot, my father marked by hundreds of them—spots of cysts growing inside his body telling the future, full of water drowning him, menace outside the windows and inside my skin—my father poisoned by the UPS deliveryman, dead in his bed. I woke up to a loud crash.

I heard pounding. "Nancy!" my father yelled in the living room. I walked down the hall, leaned on the doorjamb, and looked into the main room. A half-empty jug of Ernest and Julio Gallo wine was on the table next to the ledgers. My father stood at the bathroom door, hitting it with the heel of his hand.

"I'm going to do it, Charlie!" She screamed from inside.

"Don't! Nancy!" my father pleaded through the door.

I looked at him. His eyes were wild and fresh from sleep. I walked to him, kneeled by the gap at the door's base, trying to shrink myself, angle my head so I could peer through. I could see her bare feet on the floor inside, the lights on. She was pacing. "I can't anymore. You aren't listening to me, Charlie. Not one fucking time. I told you; this isn't fucking working. We don't have the money. You don't have the work."

"OK, OK. Just come out."

"I've got enough fucking pills, and you can't fucking stop me."

"I will call the cops."

"It'll be done before they get here."

"OK, I get it." He put his head in his hands, closed his eyes. "Just come out."

"Mom." My voice was quiet. I tried to talk under the door. I didn't feel like I could breathe. "Mom." She stood still. "Mom, come out."

"You fucking woke her up?"

"She wasn't asleep." He looked at me, quieter, said, "Eiren, honey, get to bed." I was mute, standing there, silent where there should have been rage. I went back to bed without saying anything. I listened all night. My father made his voice low. Hers was absent. I heard her slump on the floor of the bathroom by the door, crying. I heard him coax her out.

"Come on, let's sit down," he said.

I wanted him to come to my room, know I was awake, sit on the edge of my bed. But I was alone.

I could hear her open the door. I could hear them sit on the couch. I could hear her apologize by rote, flat voice, dead sound. I could hear their weight shift on the springs of the couch. I fell asleep listening to her cry. At dawn, I woke and walked past their room. They were lying on the bed in their clothes, his long legs over hers.

It took time for the Sound to become a victim of industry. It took time for it to become the principal manufacturing site for brass items—favored before the invention of plastic as fasteners and packaging and parts of industry, the river mud laced with excess copper from the smelting of brass factories. It took time before the runoff from paper and fabric mills in Massachusetts crashed the fishing industries of the coast. That copper and mercury and PCB were in the water when I was brought to the Sound as a wailing infant, in the bodies of the jingle shells of Dogshit Beach—looped through the generations of creatures. It took time for my mother to try to care for my father. It took time for her to try to care for the watershed. Both tasks, supposedly protected by legislation, made harder because the economics of valuing vulnerable lives will never balance toward profit. A bloom of something toxic, cysts growing in his body, poison coursing in the clear river, the silence of my mother, the chemicals in his blood filling him to overflowing.

Hermit crabs are liars, artists of camouflage and borrowing. They acquire and discard shells as they mature, shells abandoned by dead snails or mollusks, a periwinkle snail shell, a spiral *Calliostoma* shell, a moon snail shell, a ladyslipper. The shells are plain gray or dun. The shells are pink and swirling up to a point like a soft-serve ice cream cone, a stripe of black or dull yellow following the curves. A snail wearing the shell its body made will look level with the ground, only the protruding foot or eye stalks give it away. A hermit crab in the same covering tilts backward, and from the opening spills a tangle of legs and arms, long thin pincers, black eyes above their weaponry. The shell makes them beefy, bodybuilders with thick shoulders. But inside, they diminish, the body hooks like a thin shrimp to a pointed tail conforming to the spirals.

At Dogshit Beach, Mac resting in the cool cottage, the Sound coming back from low tide, I filled my bucket with seawater, then squatted in the tide pool, water saturating the seat of my swimsuit. I selected crabs, one by one, until there was a city of captives piled two or three layers deep. The tide finished its march away from me. I lifted the bucket. The white plastic handle with its fake braid cut into my fingers. Sunlight and colored plastic lit the crabs in orange as they scrambled over one another. They fought in my bucket, looking for a path to dominance, to escape. Unable to climb the slick sides, they would fall to the bottom as the water got hot.

I left the bucket and walked to the water, alone, even though I wasn't allowed. I walked till the flats fell away. I walked till the water deepened and darkened, until I couldn't see my feet. I felt for the sand for as long as I could, imagined a kingdom of hermit crabs below me. I floated for a second. I put my feet down again.

I could feel my heart racing. I could feel the fear behind my parents' fight, the tension Mac ignored. It was the sound of everything, the ambient noise of the summer, of something coming I couldn't name.

I took one more step into the deep. I lost track of my legs, then my hands under the water as it swallowed me further. My neck was submerged. The water was at my chin.

If I couldn't see my body, maybe I didn't have one. Maybe I'd shed it like a shell, like an anchor, like a monster. I could shed appetites and fears and hopes. I lifted my feet again and drew my knees up to my chin. My heart pounded. Vertigo swallowed my legs and arms until I couldn't feel them. I was eyes, a nose, a mouth, hair floating above the surface with nothing below it—so free I might be dead.

I closed my eyes, and the darkness spread out from below me, cold, the way sheets feel when you've wet the bed in your sleep, the warmth leaving your body and turning to chill on your skin. It pulled me into my flesh again and I gasped, sucking salt water in through my nose and choking, grasping the sand under me with limbs suddenly back to life. I stumbled for purchase, for balance, dipped my head under the water, not knowing how the tilt of my body matched gravity, then righted, water in my eyes, sand under my feet, shaking. I turned toward shore, could see the cottages, people and dogs on the sand. I found footing, ran toward the shallows, coughing water, blowing it from my nose, retching. I kneeled in the shallows and caught my breath. Close—I'd been close to drowning.

I closed my eyes and dripped. The sun began to dry the beaded water on my shoulders. Its setting cooled me again. I crawled to the warm sand where I'd left the bucket of crabs. They were riled up in water the afternoon sun had heated. I put the bucket between my knees. My hand warmed when I reached into the water. I chose a single prizefighter crab in a borrowed house of orange and gold stripes, held him as he tried to snap my fingers. My breathing slowed.

I tipped the bucket over, the warm water spilling out onto my feet. Crabs tumbled over one another, wobbling, correcting in the direction of the Sound. My fighter nipped me, and I righted the bucket and put him inside where he crawled toward the lip of the plastic and escape, reaching for something.

Atlantic Slipper Shell

(*Crepidula fornicata*)

My father found me at the edge of the tidal flats. "How'd the lesson go?"

"OK."

"That sounds like not OK."

"I can't swim like they do."

"I know. Not many people can."

"Nolan can. Quincy can." I'd seen my cousins swim with confidence and ease, their bodies mirroring Mac's.

"They can. But it is OK that you can't. You are a lot like the Caffall side. In a lot of ways. We don't swim like they do. Grandma Mac does it for her job."

"Right." I paused. "What if I'm like the Caffalls all the way?"

"What do you mean?"

"What if I get sick?"

My father was quiet. We looked out at the water for a long time. The hermit crabs had walked away. I dug a little in the sand, waiting until I got to the wet layer. I pulled an Atlantic slipper shell, a ladyslipper, out from the cool grit. I collected them, believing them to be the shoes of ocean faeries.

"It will be a long time until you have to think about that."

I remembered a story he told me about the summer he learned he had it, the summer of 1971. It was the first summer they took me to the Sound. They had just moved from New York City to the Berkshires, and they drove down to meet Mac at the cottage.

"Are you coming in?" my mother asked my father. "A swim would feel good. Your trunks are in the car."

"No, I'll hold her," he said.

"You must be hot. It's so hot out."

"I need to make sure her hernia stays in. Maybe she'll nap." He held me, tummy down like the doctors taught them, so that the hernia that kept me awake all day and all night, wailing and red-faced, would settle.

"OK. If you're sure." They stood on the covered porch of The Periwinkle.

"She's quiet. You go." No one told my parents that hernias come with PKD.

"I'll just be a minute, get changed. Then we can walk down with Mom."

"You two always swim when we get here."

"We do."

That summer was the first one when we followed the Housatonic down to Dogshit Beach, their first one as parents, the first huge red tide. That summer was the first one when they knew I had a chance of having PKD.

Our life on the Sound began a generation before, when my grandmother and grandfather and my mother's older sister, June, began spending summers there—my mother was born a few years later. And the Sound's life began long before we arrived, forged by the Laurentide Ice Sheet that covered Connecticut during the Wisconsin glaciation. The glacier pushed up the earth to create Long Island and scraped out a shallow basin below the moraine.

Erratic boulders were deposited, and when the glacier retreated and the basin filled with seawater, the river mouths were scoured out wide enough to become tidal—the Connecticut, the Bronx, the Housatonic, the Thames.

The Sound is cradled between Long Island—a narrow lobster claw of land pointing out into the Atlantic Ocean from the end of Manhattan Island—and the coastal states of Connecticut and Rhode Island. As an estuary, it is a mix of salt water and fresh, but, unusual for estuaries, it is open on both ends. To the west it joins the Hudson River; to the east, the Atlantic Ocean and the Block Island Sound. It is protected from the wild Atlantic by its massive shelter island, its water deep in the center, shallow to the edges, shaped like a humpback whale.

The Sound's salinity coursed north into the rivers for miles upstream, inviting spawning fishes, creating rich marshlands and shoals, eel grass forests, sunken meadows, fish so plentiful that before European contact, the Quinnipiac and Niantic could stand in the mud flats with torches and spear salmon drawn to the light, jack fishing. Before genocide, the population of New England was between 70,000 and 100,000 people. There may have been many more. The devastations of the early contact period—bloody removal programs, torture, burning villages, the rape and murder of women and children, introduction of killing plagues—obscures the true numbers. On the coastline lived people from the nations of Niantic, Quinnipiac, Narragansett, Montauk/Montaukett, Shinnecock, Hammonasset. On the coastline they live still.

The first dam on the Connecticut River started construction in 1794 at Turners Falls and destroyed the river's salmon run. Industry used hydroelectric power, created runoff, waste, copper and ash and paper dye, chemicals from shoe factories and cotton mills, mercury from the felting of hats, dioxin and PCBs. As easily as fish once swam up the rivers, toxicants and waste rolled down to the sea, shaped by New York City, New London, Bridgeport, the suburban sewage of millions. The population of the watershed grew exponentially, and the number of cars increased as people moved to Long Island's burgeoning suburbia, to the Hamptons, to Fairfield County. They could take commuter trains to the city at the end of long weekends, believing they'd been playing in unspoiled nature. The sea change was slow, so they dismissed it.

That first visit with me in tow in 1971, my mother and Mac changed into swimsuits in The Periwinkle.

"You ready?" my mother asked my father.

"Yeah."

"She's still asleep?"

"So far."

I had a pile of dark hair on my head, and the line from my lip to my extra-small nose made me look as if I was always trying to pucker up. They knew then what was coming for my father.

I picture them, the story of that day a story my father told me many times, about how much joy the beach gave my mother. He told me in awe of her, adoring her. She was from money, he was not. Her photograph appeared in a book published by *Playboy* in 1966, *The Classic Woman*. For a man who'd gone to night school in Jersey City with the formerly incarcerated, who dropped out of both high school and college, her parents' house in Tenafly, New Jersey, with the guest bedrooms and front lawn of ivy, was an unlooked-for gift. The beach was the same—leisure, regular vacations, things the Caffalls never once could afford. He adored her for all of that. He adored her because, despite all that, my mother and her sister were interested in counterculture and dance and experimental theater, curious and unconventional. He adored her for where she'd follow him, for her joy in the water.

"Let's go," she said in 1971, and she and Mac walked out the screen door onto the walkway to the sand. When they reached the sand, they began to run for the high tide. He walked behind them. They tried to beat each other to the Sound, running until the sand hardened into flats, and they threw themselves into the sea fast so they wouldn't mind the shock of the cold, and they felt themselves weightless in the saltwater rush, grinning as they reached their arms into the blue, pointing their strong bodies toward Long Island and ducking their heads under like seals, and cool, cool from the water after the hot car ride, light and bristling and so alive.

He walked barefoot to the edge of the waves, his toes in the water, at his feet, a ladyslipper. He never told me that he was afraid. He never told me whether he felt a new connection to me, a bonded kind of parenting wherein the child in your arms may suffer the illness that has taken the parent you loved. It is a singular feeling. When I was older, I read about Arlo Guthrie waiting to find out if he had inherited the Huntington's disease that killed his father, Woody, at fifty-five. I knew what that wait felt like. But at nine, I didn't know what it felt like yet to be a parent waiting to hear about the fate of your child.

Ladyslipper shells are small enough to hold in the center of the palm of your hand. Their shell backs appear gray or dull brown when dry, but course with stripes and smudges and flecks of yellow, pink, and darker rose when wet. Alive, they move with their domes to the sky; underneath, their soft bodies are pale cream, hidden beneath like common limpets. To eat, they send seawater through their mucus-covered gills, filtering out plankton, which they use their radula, tiny chitinous teeth, to transfer into their bodies. Alive, they pile together, older, larger females at the bottom of stacks of mollusk families. If those females suffocate underneath the weight, the smaller, younger males change sex to continue the cycle.

Their larvae have remarkable speed in the open water. They wave the long velar lobes that flank their bodies through the water, adding speed by using cilia fringing them. They can vary speeds between one body length per stroke up to four, allowing them to move quickly through water that can change in the era of collapse, extending their range and flexibility as the ocean warms, ensuring their survival, ensuring their escape north.

My parents met at Bennington College, in the theater department where she studied acting and he was on a scholarship for men to be cast in plays at the all-female college. Too poor for traditional college, he was one of only two men on campus, ten years older than all the co-eds. He clashed with professors and left early, back to New York. My mother was lost at Bennington, trailing her older sister, June, who'd been a Bennington girl first. She struggled to read and to write, thrived only in plays where the language was memorized, thrived only where her beauty opened doors. But, when her father, Doug, was diagnosed with throat cancer, my mother dropped out of college to come home and be with him. She sat at his bedside.

"Read to me," he told her. "I like it best when you read. You have the strongest reading voice." And so, she'd read as the throat cancer took his own voice from him.

When Doug died, she left for New York. She and my father ran into each other on the street, and the lure of opposites was enough intoxicant in the chaos of her mourning to see them living together in a one-room apartment on the Lower East Side in less than a year.

My father was a balm to my mother, sick with grief, Doug's ashes newly scattered in the Sound off Penfield Reef. She reveled in my father. He was tough, grasping, funny, heady. He claimed things, he claimed her, his appetite enormous, erotic, swallowing as the sea swallows, a swallowing king, a whirlpool in which she could lose herself. The disease was nowhere on their minds. They imagined a future of possibility and romance, a story my father was adept at spinning. She couldn't imagine him without his strength. He was as inevitable as the sea, vigorous in a way her father hadn't been. In a story like that, no swallowing king loses strength, his hunting days behind him; no invincible sea is broken and wasted.

After five years of living together in the East Village, building film careers, my shy mother leaning into bohemia, my actor mother married my cameraman father. Their wedding was at a country club her family paid to rent, my mother in a dress of Spanish lace she'd sewn with Mac, my father's hair still short, his goatee pointed and crisp, his tuxedo white. Their wedding party included the manager of Blondie and an Alvin Ailey dancer. My mother, so beautiful her pale skin looked sculpted out of alabaster, got high with the band around the pool. Once married, they went to Greece by getting invited on the yacht of a wealthy friend, then went to London. She wrote long letters from their cold-water flat to her psychiatrist in Manhattan, full of nightmares and thoughts of suicide, disappointment in the man she'd chosen braided with love for his fearlessness. They came back to the city, to an apartment on 10th Street with the tub in the kitchen, and finally to the forests. She trailed my father's obsession with making Shaker furniture from New York City into the New England hills, visiting the Berkshires and the banks of the Housatonic. She hovered and sheltered, weak and young, orbiting Mac, orbiting my father, orbiting the Sound.

In "Harvest of Denial," Manjoney and McKegney found that generational PKD destroyed "family cohesion and communication."

It created "time lag, availability, awareness and action-taking" and "had serious consequences in terms of continued genetic transmission, future 'burden' to the family and society, and the psychological stability of individual family members."

Two years before I was born, my parents were invited riding in a forest in the Berkshires, the woods fae and ghosted. My mother's horse was chased by a pack of dogs, spooked in a wide-open field. It ran for the trees. She tried to use her stage combat training to jump free before impact, but her foot caught in the stirrup, and she was dragged. She broke her back. She never rode another horse. She was in bed recovering for almost a year. It gave her time to think, to plan. She wanted to quit acting, to move to the country. My father wanted to make furniture in the hills. They decided to have a baby and move away.

A year later they conceived me in a room that once housed Shaker celibates, while there were still muscle relaxants from the year of spinal recovery in her blood. I was born in April 1971, just as the daffodils bloomed in New York City. The drugs slowed the bones of my skull in their knitting. The doctors told them I'd have a cleft lip, cognitive and developmental delays, the collection of symptoms referred to as hydrocephaly—water on the brain.

In September 1971, my father would burst a cyst in his kidney, pass blood for days, and in a conversation in his doctor's office, learn he was sick with PKD and a carrier as well, connected over time and blood to what came before, and to me, his infant child.

They didn't talk about it. They moved us north from a New York in full 1970s urban decay, their jobs in the movie industry in freefall. My mother took a job cooking at Alice's Restaurant. They went to Alice's full moon hash parties, smoking by the creeks that fed the Housatonic, draining bottles of wine. My mother said, "Alice knew how to have fun. And when your father took me there, I didn't even have to introduce myself. He just walked into her church with two bottles of wine, and the crowd parted for him."

One night, they brought me to a party on the Konkapot River where they smoked hash and watched the sun rise over the forested hills. When the sun was high, they walked hand in hand back to the car, and when they got there, my mother asked, "Where is the baby?"

"I thought you had her," my father said.

"I thought you did."

They ran back to the river, where I was awake in my basket and blinking at the stream. They told this story often. They made it feel like it was mine. It was theirs, and even in the haze of warm summer moonlight and hash smoke, it was a story of forgetting, time lag, family cohesion frayed.

In 1980, nine years after the year my father was diagnosed, sitting next to him at the water's edge, I watched a line of sanderlings, diminutive brown shorebirds, rushing up and rushing back, into the water, out of the water. Since 1971, the population of sanderlings has decreased 80 percent in the Americas. In groups, white and speckled gray, stick legs of black like twigs suspended above busy feet, they are cloud birds. They probe the water for food at a faster frame rate than the world, poking into the wet sand revealed by retreating waves until they find prey. They ran to the water to hunt, then ran away from the touch of waves on their feet. They ran back to the water, then away again, drawn in and pushed out, avoidant, then questing.

I ran at the question I wanted to ask my father, then ran back from it. I could taste how to ask it on my tongue, I could feel it in my feet, the walk toward it, the way to say it. But it was still out of reach, and everyone in our family answered it differently.

Patrick was shorter than his brothers, with sandy hair and a moustache and a Jersey City accent none of the rest of them had. He married young, he went into the army. He adopted a child. He leaned into the Catholicism of my father's mother and her Irish family, the only person of faith in that Caffall generation of agnostics. He fled into faith and family in answer to the question.

Brian was the most glamorous person I'd ever met. A maître d' at The Russian Tea Room, he name-dropped people he served and Manhattan acquaintances equally—Jackie Kennedy, Helen Gurley Brown, Sigourney Weaver, Meryl Streep—you could never tell what was truth and what was fabulism. It never mattered. He fled into a life he'd chosen in answer to the question. Brian wore expensive suits, drank the best wine, had a small moustache that was perfectly groomed. He had grown up with his widowed single mother after his father died when he was six and Patrick was fourteen, lived a lower-middle-class life with them in suburban New Jersey. Brian came out when he was in college, just before Stonewall, and his mother—my Irish immigrant, working class, Catholic, New Jersey grandmother, Kitty—accepted his announcement without missing a step. His identity was never a secret, even as AIDS began to turn public opinion against the gay liberation movement. What little money he got he let slip through his fingers, as if he would die tomorrow. "I'm the grasshopper, and you are the ants!" he told friends.

In the article "A Painful inheritance—Patient Perspectives in Living with Polycystic Kidney Disease," a 2015 survey of studies on the psychological impacts of PKD in family systems, Allison Tong and the other authors categorize one set of familial responses to PKD this way: "Inability to plan ahead: as the disease progressed and symptoms worsened, patients' perception of uncertainty and instability of their future also intensified. They became more conscious about their own mortality. Some resigned themselves to living day-to-day, or rearranged their goals and plans with regards to job opportunities, travel and having a family. At times, some patients made immediate decisions or seized opportunities as 'the future was too uncertain.'"

With no future to save for, he was extravagant in his generosity, especially to me, his niece. For Christmas he gave me rabbit fur mittens in pure white, a slim Yves Saint Laurent dress in shiny plaid crepe with a ruffled collar, a tiny pinky ring with an emerald. He told me, "Take care of the luxuries and the necessities will take care of themselves." He moved to Philadelphia and accumulated jobs—waiter, sommelier, interior designer. Nothing stuck. His friends called him "Lady Caffall," thought of him as a Russian ballerina in impoverished exile. The curse of PKD thrummed in the background of every choice. The curse killed him just as surely as the HIV he acquired, even though he was spared our genetic time bomb.

Denial, of environmental collapse, of death, of vulnerability, of the worth of every single life, of illness looming, is all the same. Those outside want the expert to have either a stance or a solution. Neither gets to truth. Both are waves that are the same when they are up the beach as they are when they are down, no matter what the sanderling thinks. It is the same water. It is no safer with dry feet. It is no safer with a full belly. If you run away from the question, you must always run back.

At nine—home from an argument with a child on the playground, red-faced and exhausted—I went to my father to ask what I should have said to win the fight.

"You can't change anyone's mind, really," he said.

"He didn't know anything! He didn't know about the Gulf of Tonkin Resolution."

"They think what they think."

The sanderling, once called the surf snipe, in a 1916 edition of *Auk: A Quarterly of Ornithology* is described this way: "The Surf Snipe is less shy than suspicious."

The birds are long-distance migrants, requiring staging grounds with food emerging at critical times. According to a 2023 paper in *Ornithological Applications* by Paul Smith and others, American shorebirds have been experiencing a precipitous decline. From 1980 to 2019, shorebird numbers, sanderlings included, have gone down by 40 percent. The birds must compete with human encroachment; the beaches they like are the ones we do as well. They are vulnerable to pollution and pesticides, oil spills on the ocean's edge. They compete with humans for food. In the Chesapeake Bay, we fish for horseshoe crab eggs; sanderlings need them to survive. They come back again to the places where generations before sheltered on their way around the globe. Those places have changed. We have changed them.

Since PKD is genetically dominant, most families face generations of loss, with parents fearing the future of their children. Until my generation, when many resources exist to improve disease outcomes, PKD was a death sentence, and my family expected to die young. In my father's generation, treatment was a waiting game, and then chaos. Brian and his siblings created a community of silence and shared pain as they waited to die. In a world of people paying little attention to their mortality, they were judging observers from the land of the dead. As observers, they were outside of the cycle of birth and death, and leaving behind the choice to have children, they got to stay there. But then there was my life.

In an Australian study of dialysis patients in end-stage renal disease (ESRD) from the *Clinical Journal of the American Society of Nephrology*, Ramony Chan and fellow authors found that grief and mourning were singular in those with ESRD, "broadened to comprise loss of significant elements in a person's life due to chronic illness" where "individual grief responses, rather than actual losses ... persist for many years and cause significant dysfunctions" but are "empirically and conceptually different from depression and anxiety in several studies." The grieving of lost physical integrity that kidney patients experience "is characterized by cognitive preoccupation or rumination, yearning, disbelief, stunned responses, and nonacceptance of losses...." These losses were lessened when the patient had what the authors identified as "positive affect."

Positive affect is defined by the *American Psychological Association Dictionary of Psychology* as "the internal feeling state (affect) that occurs when a goal has been attained, a source of threat has been avoided, or the individual is satisfied with the present state of affairs."

In their study, Chan and his team asked ESRD patients the following prompts, to be rated on a scale from zero at the least to three at the most extreme: *Memories of the losses upset me; I am preoccupied with thoughts of the losses; I feel myself longing for regaining what I have lost; I feel disbelief over what happened; I feel stunned or dazed over what happened.*

I want to ask my father my question and a version of these questions in 1971. He would be sitting in the doctor's office. There would be a new baby at home, the only one his entire family would produce. I'd ask him. I'd see if he would let me in. Maybe he would tell me, "every single answer is a three." And then, a sanderling, he might dash, snapping, flying, chasing waves up and down the shore, dancing toward what looked like—but wasn't—positive affect. And I might follow.

In 1971, the shell was at his feet as he held me and watched his wife and mother-in-law swim out, deep into the estuary, in cool salt water that streamed over their dark hair, in wavelets stirred up by their strong legs.

They would have my hernia fixed, still unaware of its hint at my future. The doctors would tell them that my skull was knitting together, healing. A line down my face, a strange, small nose, a line through it from the delay in my bones, would be all that remained of what was almost water on the brain.

In 1971, the shell in his hand had a back humped with a ridge like the line on my nose that had a secret in its story. The shell had a back bowed like my father's belly full of teardrops. He picked it up and turned it over. The inside was a delicate pink, tender, with a pocket that had once held the animal. It was so female, smooth as skin, color of joy.

In 1980, I held the cool, damp, gritty ladyslipper shell in my hand, then dug for another and another while I remembered his story of my first time on that beach. My father sat silent for a long time.

"You won't have to deal with it until you are an adult. Until you have kids."

"But Aunt Lois didn't have kids. And Uncle Brian didn't. And Uncle Patrick adopted Tricia."

"Well, not for a long time. And maybe there will be a pill by then. Or something."

The sun started to go down, and I put the ladyslipper into my pocket to take back to the cottage. I would stack the shells on the porch of The Periwinkle, piles of them, small noses, small bellies.

Blood Ark

(*Anadara granosa*)

"**W**hy doesn't Aunt Lois ever come here?" I asked my father as we sat on the sand at sunset, ladyslipper in my pocket, The Periwinkle behind us, my mother and Mac doing dishes.

"No one from my family does," he said.

"But Aunt June comes. And sometimes Tommy and Tupper and Rachel and Phaedra, and they stay here and have dinner."

"That's true."

"Doesn't Mom like Aunt Lois?"

"I think she does." He hesitated for a minute. "But it's easier for Lois if we go to see her."

"Why?"

"Well"—he looked out at the water— "dialysis makes it hard for her to be away from home."

"But New York is right over there." I pointed to the west, down the shore of the Sound to the Narrows where the estuary joined the complex waterways of the city, the rivers and bays and the mouth to the open ocean that surrounded the huge metropolis. "She could take a train."

"She could. But it would take a lot of time, and she's in dialysis four times a week."

"That's a lot." I got quiet. We saw Lois at Christmas. We saw her when we went to the city to visit.

"Do you want to go see her in New York sometime?"

"Could I go by myself?"

"One day you could."

It would take three more years from 1980 until I began to take the train to the city alone to see her.

Aunt Lois and Elvin lived just a few blocks from the Metropolitan Museum of Art. She worked at a Madison Avenue advertising agency, a job she'd clawed her way into from working as a secretary. She never had children. Elvin, a southerner who had grown up in Louisville, had three grown sons, spent his life recording jazz albums and teaching sound engineering. They had a tiny apartment with windows that looked out onto an air shaft.

In 1983, when I was twelve and my mother was in college, studying science like Mac wanted, getting closer to water, I visited them for my birthday.

My father had just been through surgery to make a fistula in his arm, a port for future dialysis. In 1983, it was quick surgery as an outpatient at the hospital up the Housatonic from the house over the Exxon station. They had a dialysis center—the one where he'd start his treatment—like so many regional hospitals in the decades after Section 2991 passed, like so many dialysis centers now. When they let us see him, I sat by his side on a hard vinyl-upholstered chair.

"You can touch the fistula," my father said, pointing at the bandage on his left forearm. "I'm still full of drugs; it won't hurt."

"OK."

I touched it then—the ropy arm under the bandage, the warmth of him, of his arm that had held me up and carried clam buckets and made the Shaker table where we ate dinner in Lois's apartment—and an electric shock coursed through my fingers. I pulled my hand away instantly and looked up at him, eyes wide.

"Weird, right?" he said.

"So weird." I touched his arm again. I didn't pull my hand away. The electricity was intense, a lightning bolt, an electric eel.

"It's called *thrilling*, when the blood is doubled up and goes through your arm so you can get dialysis."

When I left to see Lois, he took me to the Amtrak station in Hudson, New York—a tumbledown of hippie artists and working-class townies. Hudson formally became a port for colonizers in 1783, with sailing ships and whalers moving north from New York City on the tide of salt water, up the river valley that was a fjord, with tidal pulls and salinity coursing all the way to Poughkeepsie. By 1786 it had twenty-five whaleships, more than all of New York City. The train station had signs decorated with sperm whales, to remind you that Hudson had once drawn its wealth from the sea.

"You nervous?" my father asked.

"No."

"Well, the conductor will know you're traveling alone, and that Lois is going to get you. And you have the money I gave you for lunch?"

I nodded.

"And the dime for the phone if you get there and she's not there?"

I nodded.

"And you remember her number? And her address?"

"East 85th Street."

"East 85th Street."

"I have a book. I'll be OK."

"OK. OK. Maybe Elvin will take you to see some jazz. I know Lois got tickets for some things from work."

I rode alone down the Hudson River. Outside my window, it rolled gray in the April light. I walked to the bar car, dancing the unfamiliar rhythm of the train, hitting the doors just so, smelling the spring air between cars when they opened. I ordered my own food and paid for it and walked back to my seat with my meal and ate my tuna fish sandwich, a yellow bag of Lay's chips in the cardboard tray alongside it.

Outside the city, the train went underground, into the dark, and I saw my reflection in the glass. My hair was long and pulled back into a braid, I had braces to correct a gap between my front teeth, wore a dress with flowers all over it, country mouse.

When the train stopped, the conductor came with me to the doors, lowered the step for me, and stood next to me while I scanned the faces in the dark station platform for Lois. She emerged from behind a crowd, in a red coat she'd sewn for herself while she dialyzed, her hair cut short, emphasizing her chin, which pointed forward, efficient, confident. She held a cigarette in one hand, a purse in the other. She wore lipstick and heeled shoes, not the bohemian messiness of my mother, the rustic jeans and sneakers, the emerging gear of a geologist. Lois parted the crowd, nodded at me; the conductor looked down to see me recognize her.

"Is that your aunt?"

"Yes."

"OK, then, have a good time."

"You ready?" she asked, throwing the cigarette onto the tracks behind her, taking my suitcase in her hand, turning on her heel.

"Yeah."

"Good trip?"

"I saw the Hudson."

"That is one hell of a river." She had the long Caffall nose, the bloated face my father was getting. But there was a light in her that I hadn't seen in him for a long time. "OK, let's go. I have something to show you."

She walked ahead, her steps so brisk I ran to catch up. There were stairs up from the concrete dark of the platform, the ceiling low over us, claustrophobic, then a long hallway with marble and echoes, then another set of stairs and an escalator, then a walk under a vaulted space that opened on the huge room of the main hall of the terminal. The scale shift was overwhelming and vast, and I sucked in my breath. At first, I saw only the people, the windows pouring light into the space, the clock in the center of the room, the boards full of train names and times and tracks.

She looked down at me. "Look up, kid."

I tipped my head up into the light. The vertigo-inducing space registered as a ceiling painted a dusky, light green-blue, but then the stars, their little lightbulbs, the painted constellations, the gold and the detail revealed themselves—bright, unmistakable, celestial.

My heart opened and shimmered and all the cramped feelings of home—the ache I'd noticed first as I watched the Hudson sliding by the window, the silence beginning to peel back from me like a layer of brittle clamshell from a steamer—fell away. Underneath it was a liquid sorrow and worry that splashed onto the marble under my feet, a crashing wave that left me free and open and standing in the light reaching for Lois's hand.

"Oh."

My heart was softened by the scale, the feeling of her hand, the way she was in charge. It broke, and shining, shining in the light of it, was the possibility of everything I wanted—Mac on the sea in a sailboat or me swimming free in the ocean, no matter the depth. It was the first taste of everything.

"I know." She gripped back, looking down at me. "It's something else."

She took me home, and she and Elvin fed me and showed me the guestroom, pigeons cooing out the window and an analog clock that flipped plastic tiled numbers when the minutes changed. They played show tunes and jazz on their huge stereo. The building smelled of cooking gas and elevator carpet and cigarettes and the air shaft. She gave me a huge Walkman to use and tapes of musicals, and went to make dinner.

I lay down on the bed. The room was warm. It smelled clean, fresh sheets no one slept in regularly. I was happy. The guilt that followed happiness came crashing down so hard I started crying. I didn't even try to stop it. Like the wave that crashed in Grand Central Station, it was sudden and gutting, as if I could still touch the dark, cold house I'd left behind at home, the tension shimmering behind a membrane linking this room to those rooms. Here, on the safe side, I could feel that I never wanted to go back, and that I would have to.

"Eiren?" She came into the room, knocking softly on the half-closed door. "I heard you crying. OK if I come in?"

"Yeah? I'm sorry I'm crying. I love it here."

"Oh, I can see you do. You go ahead and cry."

She sat down next to me on the yellow-and-white bedspread, covered in a geometric pattern like all the linens in the apartment.

"Come here, put your head down." She gathered me into her lap, where I laid my head on her thigh. I couldn't remember the last time I'd done that with my mother. She didn't smooth my hair, like my mother had done when I was small, pushing it back from my forehead so softly I'd fall asleep. Instead, she put her hand on my upper arm and held it there, firm, kind. "Did you ever hear about your dad and his first time at camp?"

"Dad went to camp?" He'd never told me

"Sure, he was a Boy Scout. Eagle Scout, actually. Didn't he tell you?"

I shook my head. "He never tells me much."

"Well, he got a chance to go to camp one summer, before Daddy died. He was close to your age, I guess, yeah, twelve? I think? Anyway, it was his first time away from home and your grandpa had sent him with all this stuff, fishing gear, because, you know, they fished together all the time. . . ."

"They did?"

"Yeah, and Grandma Kitty sent him some food from home and extra clothes and everything he'd need, you know?"

"That's nice."

"Yeah, it is, of course. And, so, he was all set, and he went to change into pajamas the first night and get his toothbrush and everything." She paused. "And he'd lost the key!"

"He lost the key?"

"Yup! Couldn't find it at all!"

"So, what happened?"

"He never told anyone! And he slept in his clothes and didn't brush his teeth and didn't use any of the gear, and when your grandparents picked him up two weeks later, he was filthy and smelly and had mossy teeth!"

"Really?"

"And when they got home, he found the key and opened the trunk, and all his food was rotted!"

"Oh no!"

"And Daddy was furious. But it all worked out. He had a wonderful time at camp, even without all that stuff. It was all fine in the end. And you'll feel OK. The first night is the hardest."

"It is?" She couldn't know I was crying from relief and guilt.

"Sure. You'll see. It will feel normal after you wake up in this bed in the morning. Come on, let's have dinner."

"OK." I melted into the shelter of her, the calm. I wasn't in charge or on alert. She wasn't about to dissolve into tears or rage. She was waiting to see how I was. I had no idea how I was. It was as if no one had ever asked. And even though I'd wanted someone to ask, once they did, I had no idea what to say.

The next day and the next and the next were easier. She and I walked the American Museum of Natural History, the Bronx Zoo, the old sections of the city that showed its Dutch roots, that showed its Colonial roots.

"I'd like to live here," I told her.

"You should come for college."

"Do you think I could?"

"Sure! Your grandmother taught at Columbia, right? That could work. Or you could go to NYU. They have a good writing program."

I wondered if I could leave home, the ponds, the forests, the sky, for the city and its ceiling of painted stars lit with bulbs. She took me on the Circle Line around the island. The smell of the saltwater harbor in my nostrils, I remembered we were on the ocean. The Hudson connected me to home, even there. The city connected me to the sea.

As we floated on the Hudson, somewhere to the east of us, back in the Sound, the water was starved of oxygen, full of pollution. Even there, on the Circle Line, we were connected. Even there, on the tidal flats around the city, were blood arks, mollusks of no special notice, ordinary, but one of the few mollusks in the world to have hemoglobin in their blood, making them shimmer red, an ark to carry a creature, a color of blood in the sea. I had piles of them at home, but I'd forgotten that their bodies and blood were here, along with the blood I shared with Lois.

It would take four more years until PKD killed her. I was sixteen in 1987, the year my Aunt Lois died, and the Sound turned hypoxic and almost died itself. In some places there was no oxygen in the water at all and everything dead. It spread from a center catastrophe to the edges, toward the Atlantic to the east, toward Manhattan to the west.

Lois's aneurysm was caught early in 1987. She had brain surgery. Her body, depleted from nearly twenty years of dialysis, became infected. They couldn't find the infection. She faded, half her head shaved from the surgery, making jokes to the nurses; her husband, Elvin, sitting with her every day, holding the phone up to her ear so she could talk to me. I'd lost a good friend to cancer that year, someone who'd shared my birthday, our shared sweet sixteen party not even six months past. My mother was blacking out drunk every night.

Lois's voice was light. "You OK, kid?"

"Yeah, I'm fine I guess."

"How's school? How's writing?"

"I'm taking a class in creative writing after school, and we're going to do *The Tempest* in theater company later in the year. You should come up."

"Maybe."

"You OK?"

"Oh, kid, you know me." She sewed me dresses while she dialyzed, made all her own clothes, tailoring so you'd never guess the scars, the bulk of her organs. She worked full time. She wrote me about painting exhibitions, jazz shows, books she read. The illness may have been a curse, but in her life, it seemed an ugly dog she had crated in a spare bedroom. It had to be fed, it had to be respected, but it wasn't going to take away your life just because it was hard to control.

"Always fighting?"

"Always fighting. Come and see me. I'd like to see you. Get your parents to come. Or take the train down. Like you used to."

"I'd like that."

I didn't tell her I had a boyfriend. I didn't tell her I was trying to understand how to be loved while our family fell apart. I started dating Daniel at play rehearsal. I went skinny-dipping in dark ponds with him, to places where I couldn't see the bottom. I stood in the shallows up to my rib cage, my breasts naked in the low light, amazed I was allowed to feel alive in my skin, to own the power to disappear into pleasure. Coaxed deeper into the water, I felt the trailing reeds of murky New England hilltown ponds grasping me around my ankles. I pretended I wasn't terrified. I stayed late at the high school, reading in the graveyard with Daniel, making out at Emily Dickinson's grave. I was constantly called out from class by the school secretary, *your father is in the hospital, your uncle is dead, your grandmother had a stroke.* I was failing science and math. I followed Daniel home to his parents' townhouse, and we stayed in his room listening to Echo & the Bunnymen and the Dead Milkmen while we had sex.

A week after the call to Lois, my mother knocked on Daniel's door. We put our clothes back on and tumbled barefoot down the stairs from his bedroom. He opened the door while I stood behind him. My mother was in the doorway, pale and wide-eyed.

"Lois died, sweetie."

My hair was a mess, the house dark, the Dead Kennedys still coming from the turntable upstairs. "I just talked to her," I said.

"The infection. They couldn't stop it."

"I was going to see her. I wanted to go see her."

"Daniel, can I come in?"

"Oh, sure, Mrs. Caffall."

And my mother walked into the dark living room and sat down on the couch. She looked at my bare feet. "Get your things, we should go home."

Lois had been fighting PKD for so long that I imagined she'd always find a way to keep going. She lost her kidney function earlier than anyone in the family. She started dialysis so young that she cycled through many technologies. In 1980, Lois and Elvin and my grandmother Kitty lived for a while in Westchester, in a house they shared. In that larger space outside of the city, Lois tried home hemodialysis, a process that brings a compact version of a dialysis machine into the family home and asks the patient to designate a partner who will help with treatments. In 1980, the technology was still large and loud and clunky. Lois and Elvin dedicated a whole room to it. They disappeared into it, then he left her for five hours, from after dinner until bedtime four nights a week, while she sewed or read, the faint beeping of the machine, the whirring of the process, blood drawn in and pushed out again, monitored and quietly replicating the work that used to take place inside her body.

The technology was first available in the 1960s, and by the time Section 2991 came into law in 1972, it was gaining popularity. By 1973, more than 40 percent of end-stage renal disease (ESRD) patients dialyzed at home. Its ubiquity dropped throughout the next several decades. Now, that figure is closer to 13 percent. The reasons have more to do with people than with technology.

It is less expensive for people to dialyze at home. Machines and supplies are the same anywhere, of course, but with the work of set up, monitoring, and clean up moved into the home, a large number of people can be treated without the need for nursing or other skilled medical care. The patient and their caregiver become the medical team. This takes some training, and an outside team on call to respond to in-home emergencies. But the grinding, endless work—three to four times a week, for four to six hours a session—becomes the problem of the patient and their family or community.

A new wave of home hemodialysis is on the rise, with options like dialyzing overnight or even every single night for fewer hours. Dialysis has become a huge corporate project. Profits and a corporate center model are at the heart of the practice in a way that they were not when Lois and Elvin disappeared into a spare bedroom to set up her treatments for the night. Now, the elimination of staffing at dialysis centers makes it more financially attractive for the large corporations that run most dialysis facilities in the United States. Patients, many of whom are not prepared to operate complex medical equipment, are being talked into the work of home dialysis with the promise of freedom.

My father never used home hemodialysis. The literature on the practice cites the toll it takes on the patient's partner, the care team on the home front. The roles of patient and caregiver, already intense in a long-term, life-threatening, debilitating illness, become even more starkly drawn, even less the role of husband and wife. My parents understood that they were not equipped to do that work. Lois and Elvin were. They managed, they slipped in and out of the jobs. They spoke little about it. They were self-contained. This life they'd chosen together meant sacrifices. But even for them, it didn't last very long before the toll was too much. In a few years, Lois and Elvin gave up on it. Lois went back to a hospital dialysis center. Elvin left his role of nurse behind. They lost some freedoms, but probably gained others. It was difficult then to travel as a dialysis patient. Reciprocal care had to be organized in the place you wanted to visit, hospital paperwork, insurance coverage signed off on. It took months to plan. They stayed home. New York became their world.

"We have to go down there," my mother said, sitting on the couch in Daniel's dark living room.

"New York?"

"Yeah. Your dad says Elvin needs help with settling the house. But there isn't going to be a funeral. We'll wait until he's ready."

It took months. I was seventeen by the time he was ready. Spring made their neighborhood into a sea of flowering trees and daffodils, and we sat in the living room with Elvin where I'd listened to jazz on the stereo for so many April visits and watched my father sort through boxes while my mother drank, and Elvin used Lois's name as often as he could.

"I think you should give Eiren some of her jewelry. You know, I think Lois would want that," my mother slurred at Elvin, her eyes unsteady. She held a whisky sour she'd made in their galley kitchen.

My jaw tightened. She had no idea what Lois wanted. My mother resented Lois and feared her. I knew that. My father knew it too.

"Mom," I said, anger leaking into the single word.

My defiant tone made Elvin turn his head to look at me, then from me back to my mother. Through his southern reserve, I could read him. He knew what she was then, why I was different around her.

"Nancy," my father said, "maybe Elvin's son's wives would like that. She was their stepmother."

"Fine. Tell me I'm wrong. Fine." She was fading now, looking for a fight.

"I'm not saying you're wrong."

"Mom, he's not saying you're wrong."

"You're just as bad as he is. Who do you think you are, anyway? I'm trying to do something . . ."—she spilled some of her drink on the coffee table where Lois's jewelry box sat open—". . . nice for you. I know you loved her more than me. Right? She was like your mother?"

"Nancy!" Elvin stood up. He still held a piece of Lois's costume jewelry, a necklace of blood-red stones. "Don't you ever speak about my wife like that. Jesus, is this what things have come to?"

"What?" She looked shocked. We never spoke to her like that.

"Look at what this is doing to your daughter. Can't you see it?"

"*This?* What's *this?*" she asked. She looked stricken, about to cry, the look she had just before she broke in any conversation when she was drunk.

"That damn drink, all those damn drinks. Are you like this at home? No wonder she's not even talking to you. Lois and I can't keep her quiet when she's here. You keep drinking like this, and you are going to lose her. Completely lose her."

My mother started crying.

"I don't care if that's how you keep people from telling you the truth at home, crying like that, but this is disgraceful. You think that just because this is hard, you have an excuse to leave them behind like that. Like what you're doing? Well, you don't. You have to be better than that. They are sad too, for God's sake. Charlie just lost his sister. Eiren just lost Lois. You know what that means? Do you even think about it? Because the only person you are thinking of with that damn drink on my coffee table is you. And that will not work. It just will not work."

She put her drink down, looked at him with her chin quivering, and stood up. She walked to the guest bedroom and slammed the door.

"What the hell?" Elvin asked my father.

He shrugged.

Elvin looked at me. "Is she always like this?"

"Yeah," I said. "She is."

"Well, that has *got* to stop. Charlie? That has got to stop."

"I wish she would," he said. "I can't . . . she won't listen. And then she forgets. Work is hard right now."

"Doesn't matter," Elvin said. "Doesn't matter. No one is taking care of either of you."

Maybe it was the blood that made Elvin understand Lois and the disease, the way he hooked her up to a machine that saved her life, the way he met her after she was already a patient, his future care work understood from the beginning.

I wished I could stand with Lois in the whisper gallery of Grand Central Station, away from the vertiginous heights of the ceiling of stars, where the corridors are made so that you can stand in a tiled and arched corner far from your person in another tiled and arched corner and whisper to them and your voice can be carried past rushing commuters and sound as if they are standing beside you. I would ask her how he knew what to do in that room when her blood was a line of tubing coming out of her body, making the space shimmer with her life, the precarity of it, allowing the blood to become an ark that held her, that held Elvin too.

"You know your aunt loves you, right?" my father asked that day on the Sound in 1980. "She'd love to see you all the time."

"She would?"

"Yeah, she talks about you every time I call her. She asks about you. She thinks you're going to write books someday. She talks about going to the museum with you. Remember when we did that?"

"I *am* going to write books one day."

"I think you are too. You are a lot like her."

Jingle Shell

(*Anomia simplex*)

The last day at Dogshit Beach in 1980, I was under the cottage, reading in the cool sand, listening—floorboards creaking as my mother washed dishes, the refrigerator door closing, my father whistling as he dried off from a swim, my grandmother walking back and forth and back again as she brought towels to the clothesline and shook out the sand, never still.

I heard my mother come out of the kitchen, stand on the screen porch, stop as if watching the weather, open the screen door, step off the wooden walkway, her footsteps lost in the sand. Her face popped into my line of sight, the glow between the sand and the house, and looked into the crawl space. She spotted me. "Do you want to go on that walk to the reef? Before we have dinner?" she asked.

"Penfield? All the way down?"

"I think we have time for it."

"Daddy's making dinner?"

"Yup. So, we're off duty."

I crawled out on hands and knees in the cool, then crossed the invisible line to the warm sand of the yard. "Let me put my book away." I walked to the porch, found a sweater I'd left there, and carefully put the book on the daybed.

"Good. Let's get there before the sun sets."

I took my mother's hand. We walked to the waterline, playing sanderlings, feet wet, then dry, crossing jetties of stone, careful of the barnacles, then back to sand, Penfield ahead of us, jutting into the Sound west of Dogshit Beach. At low tide, a highway of broad, compacted sand dotted with the breathing holes of clams stretched from The Periwinkle to Penfield, interrupted by more jetties with their colonies of tucked-in mussels and barnacles and razor clams. Penfield is a set of shoals, a long curve of stones and shells, a snaking tentacle pointing at Penfield Light. You could be standing exactly on top of the reef and not notice it much. It looks not dissimilar from the tidal flats themselves. It is only a few inches to half a foot above the water, and the pale stones and shells that make it up appear as if they are a mirage or a trick of the light, depending on the time and weather. When you see it from a distance, as it passes the jetties that anchor it, the reef shimmers and is hard to catch. When you walk it, you feel as if the water of the Sound might swallow it and you and strand you in the ocean as the tide comes in, and yet you walk to the light as if you'll have some way back.

The light was erected after passenger ferries ran aground on the reef, its shallow profile visible and not visible depending on the tides. Penfield was once a peninsula so large and green that it supported grazing cows, but as the Sound has changed, it has diminished. It formed the farthest reaches of the Dogshit Beach of our imagination, the border between us and the water that led back to New York City.

In August 1945, when Mac was pregnant with my mother and their family was spending a summer weekend in view of the reef, June leapt into Mac's arms, landing on her belly, sending her into labor. Or that was the story my mother told. The early labor could have been because Mac was an older mother, forty when June was born, forty-one with my mother. My mother repeated a narrative where she was doomed to illness and delicacy from the start. She was premature, born in a hospital, then brought back to the beach tiny and unwell. When she got tonsillitis and was sent away from Mac, she had been happy, talkative, but she came home to Dogshit Beach and didn't speak for a year: every picture of her a study in rage. The displaced girl in photos from that summer aged into a young woman whose eyes shone pale gray in black and white, thin and ethereal. She told me later, when I was sixteen and trying for the first time to understand her, "My father always said, 'At least you are beautiful. You'll be all right.'"

We held hands as we walked Penfield. In the stones of the reef, there were slivers of the milky white shards of ancient bottles, newer glass pieces with the discernible label *COKE* raised on the dull green surface, sanded down to smoothness. She stopped to collect ladyslipper shells, clamshells, moon snail shells, mussel shells with their iridescent blue interiors, our pockets full. Jingle shells—sometimes called Neptune's toenails or mermaid's toenails or gold shells or saddle oysters—were our prizes on the reef.

They are a common mollusk, *Anomia simplex*, relatives to the clam, with very thin shells. They attach to marine surfaces through a byssal thread, a byssus, stretching from their interior out through a hole in the shell, anchoring the animal. As juveniles, the thread is flexible, and they traverse the soft shallows. But when it matures, *A. simplex* will crawl with its byssal foot to a hard surface, the thread calcifying as the animal holds in place. The flesh of the *A. simplex* is bitter, eaten only by the creatures like the moon snails that drill in to extract their flesh.

In the warming oceans, and ecosystems like the Sound, complex animal responses to the environmental shifts change the relationships between predator and prey. Predators like the moon snail need to consume more but lose the strength to drill into the thicker shells of some animals. Runoff and pollution in the Sound has led to significant episodes of marine hypoxia, a reduction of dissolved oxygen in the seawater. Even today, the closer one travels to New York City, the less dissolved oxygen the ocean holds. Filter feeders like *A. simplex* are less vulnerable to hypoxia, so even in the dying Sound of 1980, they were plentiful. They are listed as a species of unconcern, nowhere near endangerment in that fragile estuary, able to weather hypoxia, liberated from predation, their delicacy preserved, their breeding abundant. They are so plentiful that oystermen sprinkle jingle shells on oyster beds as a seeding substrate in which oyster embryos might grow.

When freed from the animal in death, the shells glitter. They are orange and golden and pink, nearly see-through and the size of a quarter. The holes made by their byssal threads make them perfect for stringing into necklaces for daughters or wind chimes for porches. My mother was born fourteen years after Julia Ellen Rogers wrote in *The Shell Book* in 1931, "Pretty lampshades are made by piercing the valves of jingle shells near the hinge and stringing them, then attaching the strands so as to fit over the outside of a plain glass or porcelain shade, whose brightness is pleasantly mellowed by the network of shells."

My mother's beauty shimmered around her, a glamor. Her reserve hid pathological shyness, a deep well of self-doubt, differences of mind that no one noticed. Her dyslexia undiagnosed, her obsessions and fears, habits and anxieties were left alone to run her. "No man will like it that you are so smart," she told me. "And you're not very beautiful. You'll have to pretend." To be smart was a curse around her; it meant I was a threat.

My father never told me I was beautiful. He didn't think I was. He called me *smartass* as a compliment. The slow knitting of my bones made my face unusual as a child; my nose resembled none of the noses of either side of my family. My father routinely noted that I looked nothing like my mother; my mother spoke about her years modeling. She compared me to the ugly sister in her father's family. It was a family article of faith that my looks would never be my strength. I was teased at school for the oddness of my mind, my habits, and my face. I kept my mother's portfolio of headshots under my bed, flipping through, waiting to grow into beauty, told by Uncle Brian at Christmas that I should have a nose job to make things better.

The sand on Dogshit Beach was drawn with memories of all of it, time collapsed and rewound. On the reef, my mother had scattered Doug's ashes. On the reef, she would scatter her mother's. On the reef, I would scatter hers.

When Doug died, my mother and June and Mac had walked Penfield Reef at sunset together and poured his ashes into the Sound and walked home in the rain. There are pictures of my mother and me on the jetties leading to Penfield at twilight. Seen from far away, we wait as the sun hovers over the western water. Our dog is in silhouette, trailing us, our bond a delicate thing that could break either way. If you saw the photograph, our history to that moment, nine years together, would be erased. You would not see the willful four-year-old in the rented farmhouse on the dirt road, screaming behind a locked bedroom door, "I will not take a nap."

You would not see the angry mother, back to the same locked door yelling, "You are not coming out until dinner! If you won't nap, you will sit there quietly!"

You would not see the six-year-old yelling, "It doesn't hurt!" at the mother spanking her in the living room of a rented apartment in a mill town three blocks from the Housatonic.

"You were always too much for me," she told me once when I was sixteen and she was drunk. "You're too smart. I could never handle you. You'd just look at me."

I owned the power of silence, learned from my father, to wither her, to stop her. I watched the way she set her jaw, something in it like the way toddlers look. I recognized it at four, at six, how she lost authority, became a younger version of herself in my presence, as if we were battling over toys on the playground together. Not a mother and a daughter, but peers, finally a sister she could boss. When intimacies happened between us, they were seldom parental ones.

On the reef that night she asked, "Did you hear Daddy and I fighting?"

I nodded, pulled back like a rubber band, tense to see how she would snap, prepared for anything I might have to do to get away from her.

"We don't have sex anymore," she said. "And I'm very unhappy. Your dad isn't OK. He isn't well. I don't know what I'm going to do."

She wanted me to make her feel better. There had been a time when I had done that, making daisy chain crowns in the grass outside our house when she was crying. Answering the phone for her when she was too afraid. Patting her head, hoping, hoping she would stop and turn to see my jaw locked into quiet or rage or exhaustion, turn to offer something in return.

On the reef I stared at her, my heart beating fast.

"Your father is very hard to live with. You know that. He's not OK. I'm not OK." She looked out at the water, as if she'd forgotten I was there, as if she'd forgotten I was nine. "I'm not sure what I'm going to do." I wondered if she meant about school or about being married, about whether my father would be kept or thrown away.

Some part of me formed a shell around the hurt, one that even as an adult, I would not open, one stronger than anything my mother was made of.

A jingle shell is so beautiful. A jingle shell gives protection to the animal that formed it. A jingle shell survives the rigors of the ocean. A jingle shell glitters like currency, like glass, like sunlight trapped in a shard. A jingle shell looks weak, is brittle, but survives. A jingle shell must be handled with care if you are going to string one. The holes must be wide enough to take the string. If you push, the hole will break open, and the shell will shatter and be unfit for use. There is no way to repair it. My mother seemed fit for use until you pushed too hard.

In a different family, she might have glittered in the light, a wind chime, her strength enhanced, supported, admired. It was there, her strength, but you never knew what would break it.

That year, 1980, didn't break it. The bending under the weight was palpable, but her remaining strength allowed her to enroll in an undergraduate program, discover geology, take the bus from Lee to Amherst on Sundays to spend the week at school, then take it back again Friday night. It allowed her to go with my father to surgery consultations and nephrology visits. Mac sat by her side while she chose courses, argued with my father that school was the only chance we had, made snacks for me when I came home from school.

That year, 1980, was the last with Mac whole. A series of strokes that started a decade earlier increased, angina slowed her down, and by 1982, when my mother and I came to the Sound alone to care for Mac in the cottage, it was clear she'd never be able to live alone again. My mother found a nursing facility for Mac, visited her. On those visits, every line of Mac's body showed her frustration that her muscles wouldn't do what she wanted, that her mind wouldn't do what she wanted. We were in the gas station house having dinner when the nurse called to say that Mac had choked at dinner, and that she'd given her CPR, and that it hadn't worked. She lost breath, lost air, hypoxic, anoxic.

By the time we reached the nursing home, Mac was being taken away by the coroner. I watched my mother break then, jingle shell tossed against the rocks by the action of storm.

Mac's ashes sat in a funeral home vault in Great Barrington from 1982 to 1987. She wanted my mother to scatter her. My mother, endlessly repeating the spiral of her mother's loss, couldn't do it. Scattering Mac would mean going back to the Sound. It looked like we'd moved on, but we hadn't. We sold the gas station. We moved to the college town where she'd been studying. We lived in a house we rented in the woods, the place where my father's first transplant failed. The place where he got the call that Patrick died, that his mother, Kitty, died, that he'd lost his job. It was where we lived when Lois died, and my childhood friend died, swamp down the hill, mosquitos, jugs of wine. There was a fireplace in the living room and a cabinet my father bought for his vinyl records and turntable.

There, he and I stood and argued when he was drunk, and he reached out and took my shoulders to shake me and then wrapped his hands around my throat. That moment is real, but not, like a photograph taken of us locked in it forever, my dark hair, his height. The look on his face is gone. I see only my own slim arms, a white cotton shirt I was wearing, his long pale jeans. And then it was over. I ran to my room. He never mentioned it again. There, we returned home after a trip to the Cape to see my mother's family, where my mother threw a cheese grater at my head when she was drunk and angry at me. I had no room to process anything, the losses or the damage.

At Christmas, I left the TV volume turned so low it wouldn't wake them, sitting close to read lips, the smell of Omega Oil drifting from my father's bedside where he'd massaged away the leg cramps he got from dialysis. Joni Mitchell sang on a *Thirty-something* episode. *It's coming on Christmas,* she sang, *I wish I had a river I could skate away on.* I pictured the long, frozen Hudson described in Mark Helprin's novel *Winter's Tale,* and me, on a turn-of-the-century icebreaker, moving down that fjord to the city and the Atlantic. The descriptions of Helprin's heroine, so feverish from consumption that she slept alone on the roof of her home in New York City in a tent in the winter, was with me in dreams, the grasping for breath. I opened the windows of my bedroom to the winter air and gulped in the pine-tinged cold.

I gravitated toward smart kids from stable families, the children of ministers and librarians and professors in the college town, the children of therapists. Their houses were so calm I couldn't relax. We had long-term romances. We practiced safe sex. Everyone had good hair and good teeth and was careful and kind. At night, after we baked cookies in their clean kitchens, I followed them up Mount Pollux in the dark so they could play folk music on acoustic guitars. I sang along under my breath. I didn't admit how much I needed music. I snuck out of school before theater rehearsal with Daniel, and we had hurried sex in the conservation areas behind the building, alone in woods. I followed him to the pond near his house on late summer nights, and we swam naked to the floating platform in the middle and made out, the Perseids flying overhead.

When Daniel came over for dinner, he watched my mother drink two cocktails before dinner, then three glasses of wine, slurring her words as we ate, then making herself a nightcap.

The next day at school, Daniel asked, "Is your mom always like that?"

"Like what?"

"Drunk."

I blinked at him.

"She was drunk, Eiren."

"I mean, I guess. She drinks like that every night."

"Most people don't."

"They don't?" I thought other parents hid it.

"They don't."

"OK."

"You should see the school drugs and alcohol counselor."

When I went, she asked me questions about home. I answered blandly. Just before she sent me out, she asked, "So, you think your mother might be drinking too much?"

"My boyfriend thinks so. He came over for dinner. That's why I'm here."

"How much does she drink?"

I counted up the drinks on my fingers while I listed a typical night. As I did, her eyes widened. She looked like she wanted to interrupt. She put her head down. She reached for a pamphlet.

"Have you ever heard of Alateen?"

My father closed into his illness, disappearing. He took a full-time position as a cabinetmaker and dialyzed at night, in treatment three times a week working through his frequent infections, then lost the work. He refused to take disability. Section 2991 had given him access to care, but a provision in the legislation, much like other Social Security protections, stated that the payment for care must be subsidized partly by the patient if they have any money at all. We understood, correctly or not, that certain kinds of savings were off limits from these charges—equity in a house, some kinds of investments, life insurance equity. But my father wouldn't buy a new house with the money we inherited from Mac when the Exxon station sold.

"We might move," he'd yell. "You don't know where you'll get a job."

"We're leaching money. We can't last. At least it will be safe."

But a recklessness had taken over with dialysis, an expectation that he'd be gone before the money was. Impermanence ruled him without acknowledgment. All our savings flowed away, with it, any chance of a PhD for my mother, or a house. Oil companies tried to recruit her with her fresh geophysics master's degree, but she couldn't bear the thought of working for them, no matter how good the money was.

In 1985, when I was fourteen, the call came that a kidney had been found for my father after only a few months of waiting on the transplant list. Transplants were only possible at one hospital in Massachusetts, in Boston. We moved there for weeks. While my father healed, my mother and I ate sandwiches sealed in plastic wrap from the cafeteria while we sat on the bed in the room in the hotel next to the hospital, our window overlooking an air shaft. I did homework; my mother took a leave from her job.

Kidney transplantation was still new in 1985. The technology—donation from an unrelated individual, aided by immunosuppression—was only a little more than twenty years old. The first kidney transplant to take place in the United States was done in 1950 on a forty-four-year-old woman in Illinois who had PKD. The transplant lasted only ten months. Antirejection medication hadn't been employed. It didn't exist yet. Successful transplants using immunosuppressants only began in the 1960s in America, the technology emerging alongside dialysis.

In my father's case, there was no protocol for removing enlarged and diseased PKD kidneys, so his native kidneys remained. Now, a significant number of PKD patients have either double or single nephrectomies to remove enormous or infected organs that distend their bodies and impact their blood pressure. In 1985, that was a fringe idea, dangerous and unpredictable. Most transplant patients without PKD kept their native kidneys without difficulty. The tiny healthy kidney added to a PKD patient's abdomen is dwarfed by its larger neighbors. It cannot be infected by the cysts and sits as a helpful partner to the other organs, which can sometimes serve other functions of endocrine regulation even as they fail to clean the blood. So, a third kidney, a donation from a person killed in a motorcycle accident, was added to my father's full belly. A janitor cleaning his room late one night in the hospital asked what he was there for. My father told him.

"So, you have one kidney now?" the man asked.

"No, I have three."

The man's eyes widened, then his face broke into a grin, and he began to laugh. "Three kidneys!" He cried, doubled over, "Three kidneys!" Every time he passed my father's door for weeks after, he would lean in and point. "Three kidneys!"

In the years around that first transplant, the losses came so quick and heavy that it was hard to breathe. Before my father had his surgery, my Uncle Patrick died on the table during his own transplant surgery. Before his transplant, Uncle Patrick had had a stroke while on dialysis. His dialyzing took place constantly inside his own body through something called peritoneal dialysis (PD), as a line connected his internal organs to a bag of chemical cleaners strapped to his belly. These he changed every day. They came in a suitcase so that he could travel with them. His claustrophobia in the dialysis center, in the dialysis chair, was so intense that he'd contemplated suicide. PD was the only thing keeping him tethered to a daily routine. While on PD, he had the stroke that took the use of his legs. Angry and depressed, he came to Christmas in his wheelchair, talking about the transplant as if it would change everything. But the transplant would take his life.

We all feared that the transplant might take my father's life as well. Uncle Brian took the train up to Boston to see my father in the hospital, brought me sweaters he'd bought in Paris, pulling French chocolates out of the pocket of his coat. Brian knew, like I did, that dialysis was a grind—the diet restrictions, the hours on the machines, the weigh-ins, the thirst, the pain, the deep purple bruises like tattoos from wrist to elbow, the leg cramps, the outline of the fistula against his skin. We all hoped that the transplant would ease that.

A month after the transplant, we came home, installed my father on the couch to recover. He started running a fever before Christmas. In two days, he was delirious and manic, refused to see a doctor, and by the third we drove him to the emergency room. He lost the kidney.

On Penfield Reef in 1987, three years after he'd lost it, dying water around us, we were alone with Mac's ashes. We'd driven down the Connecticut River. We were all very tired. It was the last time we would walk the reef together, Mac's ashes in a metal can inside a plain brown box we'd taken from the funeral home. I blinked and imagined Mac dying six years before, choking to death in her nursing home, gasping for air. In a human body, hypoxia can result from chronic conditions like asthma or heart disease, from a virus like COVID, from anemia or cyanide poisoning, from choking.

All around the reef, the Sound was dying, hypoxic, exhausted by years of red tides, runoff, neglect. Marine hypoxia is most common in shallow bodies of water like estuaries. When it occurs, dissolved oxygen levels drop, usually from an overwhelm of nutrients flooding an ecosystem—nitrogen, phosphorus, imbalanced saline—the same things that overwhelm a body when kidneys no longer filter blood properly. The roots are the same as those that cause HABs —runoff, soil erosion, sewage release, deposited atmospheric nitrogen. A healthy coastal ecosystem will have 5 milligrams of oxygen or more per liter of seawater, even up to 8 milligrams per liter. A hypoxic ecosystem will have half that, under 4 milligrams, but usually between 2 to 3 milligrams of oxygen per liter. Anything under .02 milligram is considered anoxic.

In 1987, after decades of runoff and sewage entering the water, the oxygen levels dropped to zero in some places in the Sound, and the animals began to suffocate. The entire lobster catch was lost. Lobsters suffocated in holding pens where they waited to be shipped to restaurants, they suffocated in traps. Some animals tried to move toward water that had more dissolved oxygen present. Somewhere in the dark, the summer was coming, the cysts of *A. fundyense* on the bottom of the Sound.

By the third week of July 1987, the bloom of algae came. As Andersen wrote in *This Fine Piece of Water*, the arrival of destruction was recognized by a researcher on the Hempstead Sill, who saw a brown flood spread slowly through the blue water, choking visibility and life: "It was not sewage, as swimmer and sailors who encountered this brown pall assumed. It was algae, a big-bang universe of tiny one-celled plants, reproducing, growing, spreading across the Western third of the Sound like some out-of-control science fiction creature" covering dozens of square miles. First it was born, then it grew, and at last it died, the whole blanket of it expiring at once, sinking until the murky water cleared to the crystal blue it had worn in the time before colonization, eerily clear. At the center of this unnatural seascape, a core of water with no oxygen at all, a measurement of zero milligrams per liter that made the researcher doubt her equipment.

The hypoxic water created a chemical reaction that released hydrogen sulfide from the sediments at the Sound's depths, a rotten egg smell hovering over the Sound. The hypoxia spread to the shore, to the harbors. The list of fish lost to hypoxia was enormous—summer flounder, blackfish, horseshoe crabs, eels. The commercial implications to fishermen were devastating, especially for lobstermen. But the consequence was also felt among fish that form the necessary midline of the food chain. Among those losses were menhaden, fish that made the estuary home, made it a nursery, and had experienced it as a killing ground. In 1987, though most of the water was low in dissolved oxygen, a portion of the Sound became anoxic.

We stood where we'd walked when I was nine, where the jingle shells sat in heaps on the coarse sand, where I'd known, first and fully, that PKD was breaking us, and that we were not well. We didn't know how to be well. We would always be looking for how to be well.

Soon, my mother would quit drinking. Soon, I would go to college. Soon, I would leave college so that my father could have another transplant. That one would take. We didn't know that the first transplant had left an infection that would kill him by sixty-four. We didn't know that only seven years from that year when the Sound almost died, my mother would divorce my father, and I would become his caretaker.

"Should we scatter her?" my mother asked.

"We drove all the way down," my father said. The light was weak, an early summer weekend where the air was still cool, and the summer people hadn't arrived.

"I know, but is it legal?"

"No. It isn't legal. And it wasn't legal with your dad, right?"

"I feel like June should be here."

"But she's not."

My mother was silent for a while. "Should we say something?"

"We should say something about her," my father said. "She had a big life. Lots of adventure. She loved to sail. And she was always reading. And she loved swimming." He looked at me.

I had been silent the whole way down. I didn't want to talk about Mac. I loved her. But the beach reminded me of how she held me in the water, like she knew I'd fail, like she suspected I'd never learn to swim through the ocean or through my own life, like she was desperately hoping I'd prove her wrong, like I was too much like my dad.

"She taught me to swim," I said.

"That's right," Mom said. "She did. She taught you to swim. She taught me to swim and to sail. And she took care of us." She started to cry.

My heart went cold to her. I remembered a fight we'd had the week before,

"You little fucking bitch," she had said, "you're lying. You didn't tell me you needed a ride home from rehearsal. You never fucking tell me anything." I'd closed the door to my room while she stood outside it and screamed, "You're a fucking liar."

"She really loved it here," my father said.

"She did," my mother said.

"We should scatter the ashes, Nancy. There's no one around."

My mother nodded, then walked to the edge of the reef. She opened the box and put it on the sand, then took out the metal container and took off the lid. She walked to the water and poured the ashes in, watching the waves move over them, watching them wash out into the Sound, watching them wash back onto the shore.

There are movies of my mother and me on Dogshit Beach— the dog chases a stick, we stand against the sunset, I wear a swimsuit bottom and plastic sunglasses, I hold a peach, I hand it off, an adult hand comes into frame to take it, it is hers, I know the thin fingers with the gold ring and its black intaglio of the griffin. I run to the water, I have a toy, a troll doll, I sit in a sweater and pants at the edge of the surf, my hair is toddler short, my diapered bottom is in the water, I dip my doll into the shallow pool made by a tire, pretending it can swim. I am older, I am sucking seawater from the end of my braid, my mother gestures to take it out of my mouth. I am one, I am four, I am six. I am a water baby who cannot swim, who fears the depths of the sea, who never tells, in pain, afraid, and lonely. We are walking to Penfield Reef. We are always walking to Penfield Reef.

On Penfield Reef in 1980, it began to rain, the drops falling on my terry cloth robe, dampening her hair. She looked at me like I might reply about her marriage, the sex, the failures of my father. I looked at the shells I'd piled up. I started to put them into my pockets. She shook off the intimacy I failed to return.

"We should go back, honey," she said. If she remembered telling me I might inherit my father's disease, she didn't say anything. She didn't see me there, all those moments of me and the water, struggling, layered with meaning. The history of all of us stretched back into the sand, the memory of the water, down into my mother's first summer on the Sound, forward into the time I'd scatter her on Penfield Reef. And nowhere did she see the girl in the dark, afraid of the black spot, afraid of the deep water away from the shore. She stood with me, next to me, and didn't know how alone I was, how I'd float away from her for the rest of our lives.

"There's a pile of them." I pointed to another collection of jingle shells, showing where they lay, glittering as the light changed, a pocket of sunset tucked into dull rocks and clamshells on the reef.

"They'll keep," she said. "There'll be more of them when we come back."

She took my hand. We walked along the flats, past the longneck clam holes, by the jingle shells; around us in the water were dinoflagellates, moon snails, long-armed hermit crabs, blood arks, slipper shells, and so many more that I couldn't count or know them all—menhaden and alewives, oysters and mussels, salmon and trout, sanderlings and eels, jellyfish and sea stars, and somewhere, far from the Sound where they once had swum, humpback whales and their calves, parents and children, generations of them.

By the time we stood in the sand front yard, the clouds looked heavy, and my father was closing up the grill in the rain.

"Dinner?" he asked her.

We sat on the porch, wind blowing water through the screens, plates of summer corn in front of us, salad and burgers, Mac on the picnic table bench across from me, her tanned legs tangled up, her bare feet making sandpaper sounds on the sand-dusted wood floor. After dinner, I strung my jingle shells in cascades. A cold wind blew in from the summer storm, pushing their chimes—a tinkling, brittle echo in the evening gray, in the oncoming dark.

SEA OF STARS, 1994

And men go abroad to admire the heights
of mountains, the mighty billows of the
sea, the broad tides of rivers, the compass
of the ocean, and the circuits of the stars,
and pass themselves by . . .
—SAINT AUGUSTINE

Dinoflagellate Microalgae

(various species)

I was twenty-two years old that July of 1993 when the call came. I stood in my parents' kitchen talking to my father, and the telephone rang and I answered it and there was the news. My father's nephrologist had only given me an ultrasound to rule out PKD. Dr. B didn't expect to find anything. I was too young.

"This is Dr. B."

It's Dr. B, I mouthed to my father, who was sitting at the kitchen table. He nodded, his body rigid.

"I should have made an office appointment for you," Dr. B said, "but I thought you'd be fine."

I couldn't catch up to the sentence. I stared at the electric stove in the corner, the pottery jar full of kitchen tools, the antique wooden mallet, scarred and stained.

I walked to the picture window until the cord on the phone pulled taut, the limit of my distance from my father. I stared at the hill that the English settlers of Massachusetts called East Mountain. Beneath that ridge, the Wôbanakiak: Amiskwôlowôkoiak, "People of the Beavertail Hill," made their village. Ten thousand years ago, that hill was a beaver-shaped island swimming in glacial Lake Hitchcock. As the waters receded, the Wôbanakiak: Amiskwôlowôkoiak told a story of a giant beaver decapitated and left to form the low mountains of the ridge. I looked at Beavertail Hill, phone in my hand, and I knew that the water of my body held toxic runoff. A healthy future receded from me like a tide, and I reeled with vertigo. All at once, I fell into an oceanic whirlpool and watched myself being pulled down, as if a camera zoomed in on my face while the background fell away and blurred.

For the Fourth of July, just a week before the call from Dr. B, Ben, the boyfriend who had loved me since the summer before I started college, took me to stay at his family's cabins on the Atlantic. His father and stepmother had taken us out on their boat, sailing far enough from the cove that the wind changed, the coastal smell of the shore replaced with a deeper, older scent. After the sun went down, our wake burned a luminous bluegreen— bioluminescence, shining when the waves moved it, tingeing our throats, the undersides of our cheekbones.

I'd seen bioluminescence before—the same bluegreen—while swimming with my mother and Mac and June in the Sound in 1980, when June and her family had come to visit. Late at night, after dinner and dessert, we left the men in the cottage and took towels out to the dark beach where the tide was high. We peeled off all our clothes and tossed them down in piles and looked to see if anyone could see us there in the dark. We ran to the water as fast as we could, holding hands, screaming at the cold air, screaming as we hit the cold water, the waves covering us. I'd never been naked in the ocean, in the dark, and the water was warmer than the air, salty and buoyant. Around us the water lit up.

That moment on Ben's parents' sailboat returned to me as I held the phone in my hand—the swirl of it, its fire and peace, confusing and healing and otherworldly. The bioluminescence both took me out of my body and into my body. I remembered it as the whirlpool drew me down.

"You should come in," Dr. B said over the phone. My body had disappeared, but I could feel my father behind me, his eyes on my back. "Come in on Monday, all right? We'll talk about it. I'm sorry. Tell your dad I'm sorry."

In dark places, it is light that calls us. Even in a broken ocean full of chemical fouling and warming water, bioluminescent fire remains. Sometimes, a warming and failing ocean makes it burn brighter. Spanish colonizers, naval officers tracking U-boats, and generations of sailors have seen this glow. We've written about bioluminescence for as long as we've kept natural history records, both as a phenomenon of mystery, and as a support to human life.

Bioluminescence is a chemical reaction, usually produced by the combination of luciferin, a compound, and luciferase, an enzyme. It creates a glow within the body of an organism or outside of it in water when the chemicals are ejected. The bioluminescence I'd seen in the Atlantic was caused by dinoflagellates, a phytoplankton. They can appear in freshwater, but the majority, and the only ones that glow, are marine. A dinoflagellate will photosynthesize, making energy from the sun like a good floating plant. But they also synthesize luciferin and create luciferase. Their luciferase is related to chlorophyll. Their bioluminescence produces no heat, a cold fire in water.

Dinoflagellates are essential to the marine ecosystem, a bedrock of the food chain, a marine alga. Dinoflagellates who produce bioluminescence are often found in bays or estuaries, concentrated into narrow confines, creating a density in the water. In this density, they glow when disturbed. A 2017 study found that this bioluminescence can reduce the grazing of copepods on dinoflagellates, making their glow a protection against predation. Every stroke of a swimmer, dip of a canoe paddle, flash of a wave, wake of a boat creates a nexus of light. Often, in the Atlantic, this fire is a pale blue, the light of stars, the Milky Way.

And that light is sibling to other cold fires, so called because only 20 percent of the light emitted produces heat. Before the invention of safety lanterns, coal miners stashed fireflies in jars and took them underground, lights that would not explode in the deep pockets of the earth. In World War II, Japanese soldiers carried dried plankton, revived them with water, and, in their glow, read maps and drew no enemy fire. In the Indian jungle, travelers navigated by glowing fungus on the trees. To find your way where a simple candle might cause the air to explode, where the path is too dark, where light could summon danger to your doorstep, you will follow any light. But bioluminescent light, created without fuel, reproducible, naturally available, is holy.

On the day Dr. B called and told me the news, Ben drove us to the top of the Holyoke Range, a line of east-west marching basalt that defines the southern border of the inland New England valley where my parents lived. The ridge of the mountains was formed 200 million years before, as riffles of lava emerged and built, and as it weathered, pillars of rock emerged, giving the hills a rusty color. He gave me a box of blueberries. We looked down on the Connecticut River winding to the Long Island Sound—my childhood estuary full of dinoflagellates—past tobacco fields, factories.

"I think it'll be OK," he said. "You've got time. Just look at your dad."

As he spoke, I stretched away from him. My father was on his second transplant, surviving longer than anyone else in his family. Ben hadn't been there for the first transplant, the deaths of my uncle and my aunt, the poverty, the terror, the chaos, my parents' brittle marriage.

Ben's mother died young. She said goodbye to him from the bed where she'd been since after the chemo stopped, just before he went out to the movies. She said, "I'm of a more Puritan soul," thinking he was breaking rules, going out instead of going to school, unaware it was a Saturday. When he came home, she was gone.

Her Puritan soul was on my mind. I would die young. All loyalty and goodness, he would stay with me. He'd go through all of it again.

"You remember when your mother came for her shoes?" Ben's father, Big Ben, asked him the night we sailed the ocean of stars.

"You mean when her ghost did?" Ben laughed.

"They washed right up next to us on the shore. We lost them after she died, you know? They washed off the boat," Big Ben said to me as he took a drink from the cap of a thermos filled with martinis. "And then, there they were, on the shore." He looked at Ben. "Your mother came for her shoes."

I told myself that Ben's mother's death was why, from the moment I hung up the phone with Dr. B, I wanted to break up with him, like a hunger, down in my body, calculating and eerie.

I told myself I was protecting him. But I was protecting me.

I wasn't the girl he'd fallen in love with, who could have children and make a life with a summer house and a sailboat. I couldn't stand how he looked at me on the top of that volcano, the same look my mother wore every day of my childhood, bound by love and duty to a fate that wasn't in her body but in my father's. She looked at him like a porcelain-cased bomb that could explode at any time. I'd seen her look at rivers the same way. She knew how sick healthy-looking things could be. Now she knew the same of me, and so did Ben. I wanted to get as far away from that look as I could.

Time has changed how humans use bioluminescence. On the island of Vieques, Puerto Rico, there is a famous bioluminescent bay, Puerto Mosquito, named for a legendary pirate ship, and not for the insect; you can no longer swim in the bay. Swimming tourists and their sunscreen and DEET bug repellent have diminished its fire. At one point, the *Guinness World Records* named it the most bioluminescent bay on Earth. Its native dinoflagellate, *Pyrodinium bahamense*, was found in its highest concentration at 3.3 million organisms per gallon (876,953 per liter) of water in March in the wake of Hurricane Maria in 2017. Tour guides tell a story that the bay was corralled by the colonizing Spanish, who thought the light was the work of the devil. When they tried damming the glowing water to contain its danger, the increased concentration of plankton only made the bay glow brighter. When Hurricane Maria devastated the mangrove forests that protected the bay, the influx of water diluted the concentration of bioluminescent organisms and washed the dinoflagellates out to sea. For a time, there was little to no bioluminescence to be seen. As the island and the bay healed from the storm, the blue fire returned.

In 2018, a swirling, glowing whirlpool was detected by satellite in the Baltic Sea, so huge that it could contain Manhattan within its circumference. The influx of bioluminescent plankton multiplying in those heating waters were also turning them more hypoxic than they have been in 1,500 years. This bioluminescence indicates the illness of the sea, its movement toward collapse. In India that same summer, and for several summers after, overabundant algae glowed blue in the waters off Mumbai and Chennai, attracting tourists who came for the light, unaware that it indicated a dying ocean whose warmth encouraged deadly algal blooms. Scientists believe that warmer waters globally will make the oceans more bioluminescent as plankton respond with increased breeding.

In his office, Dr. B said, "I wish I could tell you some better news. That many cysts for someone of your age? It is pretty advanced."

"The family story is that it's harder on women."

"I don't know about that."

"My aunt was twenty-four when she went on dialysis."

"I think you'll have five years of good health before you lose kidney function, go on dialysis, go through menopause. You'll never have children. Your life expectancy is similar to hers."

"So. I'll be twenty-seven when I lose function. And forty-nine when I die?" I got quiet. "Lois died of a brain aneurysm at forty-nine."

"You've got an 80 percent higher chance of those than the general population."

"OK."

"This isn't what I wanted to be telling you."

My family drowned under the news; my father started drinking again. We went to dinner together. Sitting at the picnic table, the Holyoke Range in the distance, a roll of paper towels flapping in the wind, a glass tumbler of red wine in his hand, he said, "I'm not my PKD. You aren't either. I'm more than this thing."

"I don't know. I don't think I can separate it out."

"If you don't, it'll kill you."

My mother cried when she looked at me. She'd dress for work, spend the day testing groundwater, come home, hide in her bedroom reading sci-fi until she slept. Standing next to my parked car, summer darkness and the sounds of crickets surrounding the farmhouse, I'd see her light still on, and the Milky Way coursing over the house like swirling plankton.

I drove aimlessly up and down the coast, and the water let me know what to do. A few weeks before we left for my last year of college in Seattle, I told Ben we had to break up, and wept in bed afterward, sure I'd made a mistake from which I'd never recover.

Back in Seattle, I didn't call my parents, stopped speaking to friends, wrote a thesis on dying women, drove north to the Puget Sound and Deception Pass. I knew the Pacific was under enormous, generational, and existential threats, but I craved the euphoria of an animal surfacing from the gray-black water—a whale back, a seal's blocky head. I had no idea how to live with the grief for my body or for them, twinned and heavy. By the time I graduated the following summer, Ben was in love with someone new, and she was sweet and healthy and studying geology. In the picture Ben showed me, she looked like someone you could have a sailboat with.

I thought if I went back to where the ocean smelled right, the waters glowed, I might find an answer. I bought a cheap car and packed it with books and a tent, and left the West Coast, singing sad songs with the windows open with Heather, the new friend I made in a seminar on the sublime. Heather, who couldn't drive, was the first friend I made in my new body, standing in the back of the Re-Bar, our favorite bar, seeing bands, transformed by music while I nursed a ginger ale, pretending I could make PKD go away by refusing sex, joy, alcohol, food. Her melancholy and separateness from people felt familiar. She handed me cassette tapes as I drove, smoked cigarettes with red lips and effortless cool, reminded me I was raised by grieving bohemians, and had been admitted to the club of people who expect to die young.

No longer just a light in a jar, bioluminescence is used now to test water quality and to indicate overheated oceans, ecosystems damaged by rising seas. We use it for health diagnostics and to research renewable energy sources. One of the first medical uses suggested for bioluminescence came from Roman naturalist Pliny the Elder in his 77 CE work *Historia Naturalis*. He wrote that a bioluminescent jellyfish we now call "night light of the sea," *Pelagia noctiluca*, could be boiled in water to drink or added to wine and used to treat kidney stones and gout, both companions of PKD.

After the bloody Battle of Shiloh in the American Civil War, on a field littered with the bodies of injured men, there in the dark, some of their wounds became illuminated with a pale fire. The men whose wounds glowed recovered faster. Nurses and doctors at the scene called the phenomenon "Angel's Glow." In 2001 it was reported that teen Civil War buffs working with scientists from the US Department of Agriculture learned nematodes in the battlefield's soil mixed with blood in the wounds, feeding bacteria until it glowed. That fire's antibiotic elements helped heal the soldiers. This same bioluminescence could also be used to protect agricultural crops from the potato beetle. A naturally occurring bioluminescent bacteria, *Aliivibrio fischeri*, occurs symbiotically with Hawaiian bobtail squid. When submerged in water with high levels of toxicity, the respiratory functions of its bacteria are interrupted, indicating a polluted waterway that causes their light to wink out. It is hard not to ask the pale fire, and the entities that create it, to become our servants as we crave their magic, as they warn us, educate us, and protect us.

When I was diagnosed in 1993, I was told there was no cure, but every spark is a clue, every shimmer in the black ocean a way to something more luminous and complex. In 1953, Jim Lovell— the savior of the Apollo 13 mission in 1970, who would navigate to the moon twice—was flying over the Pacific Ocean when his navigational gear failed. In the lowering twilight, he had to find the aircraft carrier, and he needed something to steer by. As he approached the carrier, the glow of bioluminescent algae, stirred up by the boat's propellers, guided him home.

When we graduated, Heather and I drove east from Seattle, pointing to the Atlantic. We stopped in national parks and a mission church in Idaho that ached with echoes of genocide. I was permeable to every broken thing, the wound of the West, hollowed out of animals and people and replaced with an empty simulacrum of place, or a place that was so obscured and complex I couldn't read it at all.

We slept in motels. When Heather was in the bath, I'd lie on my bed and close my eyes, the whir of the road in my ears. I'd fall into a trance, like descriptions I'd read of deep meditation. Buffeted by fear, I wasn't in control. I lacked context for the mystery trying to find me, which felt bottomless, another whirlpool.

South Dakota rolled like waves, humming with the water it remembered, grass shifting as if in tidal pull and not in wind. We camped in the dry heat of the Badlands in a campground called Sage Creek, and there was nothing to distract me from my feelings. The first night, we sat under the stars, both thinking about death in our own ways. Heather lived with depression, suicidal ideation, and poverty all her life. We drank bottles of screw-top wine from a six-pack we bought at Wall Drug as trees swayed in the dry wash to our east, no campfire because of the drought.

"I don't ever want to leave here," she said. The first either of us had spoken in hours.

"I know. Me either." Without fire, we had nowhere to turn our focus; the dry land all around felt too large, vertiginous. But the pull of the fire of the stars overhead, shuddering behind wisps of clouds, drew our gaze as the flames might have. They wheeled over us, a time-lapse image. It was what I had been avoiding since diagnosis—the edge of the whirlpool.

"Sing something."

I sang Vic Chesnutt's "Dodge," his lyrics about leaving home echoing in the stillness: *I done shit everywhere that there is to eat/ Guess it's time for me to get the fuck out of Dodge.*

I remembered Dr. B's office—*This isn't what I wanted to be telling you.*

In the silence after I sang, I thought I heard a bison snuffle. The darkness was near absolute, the sky overhead a rush of clouds. Heather took her eyes off it to look at me. "This is the first place I've been where I didn't want to kill myself."

"This is the first place I've been where I thought it might be OK if I died," I replied.

She handed me another bottle of wine; I cracked it and leaned back onto a blanket laid over the hard earth and watched the speeding clouds until I was too tired to watch any longer.

In the morning, the ground was alive with grasshoppers, a wave of them kicked up by every stride. They shushed out ahead of my feet, a bloom of insects born of drought, and I walked to a dry wash surrounded by brittle cottonwoods. They made the path an ocean, almost luminous, waves of grasshoppers on the dry prairie, like a flush of bioluminescent fireflies in the grass of a home meadow, the flash of dinoflagellates as I splashed into the night-time ocean of the Sound. On the horizon, a herd of bison grazed. That night, I went to sleep without watching the stars, exhausted by the holiness of the place. I dreamed that we were in a sailing ship on the ocean, hunting whales—I was a sailor, Heather was a sailor, we'd always known each other over lifetimes.

A windstorm woke me up after midnight. The sides of the tent flattened over us in the gusts. I came to, unable to breathe, nylon covering my mouth. Heather slept on, and I unzipped the door and crawled sideways out into the cloudless South Dakota night, nearly knocked over by the wind. Randomly, in the full dark, I pulled at the sides of the tent, lacing an extra rope through the fly hooks on the top to keep it from blowing sideways, tying it to a windbreak picnic table we'd camped next to, pounding the tent pegs back into the packed earth. When the tent was steady, I looked up for the first time and saw the Milky Way stretching like a backbone over my head, listened alone in that wind to the sound of the distant trees, the whickering of horses from another camper's trailer. The universe was so much bigger than the fears I was carrying.

I thought, *What if the whirlpool I'm falling into is really a vortex that takes me out through the glowing oceans of the world and into the infinity of the stars? What if that's death, something bigger and wilder and more beautiful than the fear?*

I closed my eyes then, standing there, and felt exhausted, as if I was coming in from a long hike in the rain. All I wanted was sleep. The dead sea, the ocean of stars, the sea of grass, the memory of sailing—face illuminated blue with bioluminescence, ocean alight and alive—buzzed through me with something like an answer.

Book Two

GULF OF MAINE, 2014

—

It is advisable to look from the tide pool
to the stars, and then back to the
tide pool again.

—JOHN STEINBECK,
THE LOG FROM THE SEA OF CORTEZ

Northern Rock Barnacle

(*Semibalanus balanoides*)

On the night of the nor'easter, inside the Coast Guard cutter, it was 2014, but it was also 2008 and 1980 and 1987. I was caught in the spiral of crisis, so disoriented I couldn't separate time. I struggled to breathe, anoxic. I could see around me into the low darkness of the tiny space—the faces of the Coast Guardsmen, Dex swallowed in his orange dry suit. All the seats had double seat belt straps, like a child's car seat, and we were so deep inside the water that even stepping aboard the craft I was drowned in a wave of seasickness. They gave us pills to dull the nausea, bundled us in like toddlers, and Dex fell immediately into a self-protective sleep. Later he would tell me, "I dreamed about a drawing of a flower."

We were rolling over the Gulf, in the midst of its fury and chaos, caught in a storm the warming water fueled, in the teeth of change. I let my mind fade in and out, dulling my thoughts, trying to give in to the power of the water. I tried not to wonder if the accident, the collapse, the seizure, was part of my health beginning to fail, the long, terrible slide into the worst things. I closed my eyes and rode it, that storm; my child, across the belly of the ship, rode it too.

The waves became a musical score, a set of drumbeats or heartbeats or the roars of sea creatures imagined by sailors. It was a holy roaring. The waves had information. They were the depths I had feared since childhood, the danger I was in as a single mother with an incurable disease. They were every life I lived before; sailing ships carried me through rain-whipped air and into the heart of a swallowing mystery.

It was hard not to let thoughts grow in the face of the roaring. It was hard not to fear that the boat would capsize. The seas were the largest I had ever been on. Even the worst ferry trips to Monhegan, crossings after storms where swells took the *Laura B* broadside—were light compared to the nor'easter. It tossed us, and the vertigo of the accident was dwarfed by the disorientation of the waves. We were sideways and pointed to the sky. We were pointed to what felt like down, but there were no windows in the hold and no way to tell, after a few minutes of tossing, which way was up any longer.

"Hang on, miss," said the Coast Guardsman as we peaked on a wave, held for an instant, then fell in a cartoonish return of gravity.

OK, I mouthed. My voice came out a croak.

He tried to smile at me. "It's bad, but it's short."

"OK."

"And we'll be there soon."

My mind flicked in and out. There was something I needed to know in the waves, something in that most profound kind of whirlpool. If only I could listen and not sleep. If only I could remember what I needed. If only I could tell if I had destroyed everything by trying to come to the island at all.

I came back to Monhegan again and again to avoid darkness. I came for wonder. I grasped it from loss, learned to hold it, even for a moment, beside all I feared—a practice, a muscle I was building, a muscle I would need.

In 2008, I brought Dex to Monhegan for the first time, knowing he needed the muscle too. He was three and a half, and I was thirty-seven. We rode the *Laura B* to the tiny harbor and stepped out onto a dusty road that led to the village and found ourselves together on an island in the Gulf that I had only been to a handful of times, and that he, a skinny toddler with a broken arm, had never seen.

I took a photograph of us that first night on Monhegan—six years before the nor'easter, two weeks after I left his father, fourteen years after the whirlpool of stars.

In the photograph, I held him, facing the camera. His back was to it, his blond hair gold in the late afternoon light, one arm around my neck, the other in a pale blue cast, the exhaustion curving his shoulders, an overload of wonder or of sorrow. I took the picture with my good camera, holding it, holding him. He smelled of salt water and seaweed drying on his shorts, heavy with the gathering density of a three-year-old moving toward sleep. My body listed. The work of leaving his father was in the set of my shoulders. I looked at the lens, trying to convince it I mattered. I looked as healthy as I wanted to be. I did not know I was beautiful. I wanted to send this picture home and save my marriage.

I married Jonah because I loved him, and because he looked safe. I married Jonah because my father had just died, and because I was terrified. I tried to pass, as middle-class mother, as well person. But Jonah wasn't safe. He lost his job in the 2008 financial crash. Our health insurance went with it. I was uninsurable. Panic became a flow of mud through the house. The real estate market was tipping over; we were nearly underwater. Dex broke his arm and bills arrived and we unraveled. I had not lost my kidney function. I was as old as my father had been when I was born, the year he was diagnosed. Dex didn't sleep like other children, he spoke early, he walked without crawling, sensitive and alert and just like my mother's stories of me. He took all my energy, and I could feel my kidneys changing. If I talked about any of it, Jonah would flare and bloom to fury.

Jonah's rage was a weather system I learned to predict, amateur meteorologist. I avoided him, leaving the house. I gave up talking about bills. I gave up talking about my health. Nights were full of arguments. In the daytime I called my mother to beg her for a plane ticket.

"Is it that bad?" I remembered her behind the bathroom door. I wasn't sure I even knew what bad was.

"I don't know."

"Go for a walk, maybe it will blow over." As if she also understood Jonah as a storm.

On Hartigan Beach near our apartment, the densely packed buildings of Rogers Park and dorms for Loyola University fell away with a sudden drop at the edge of Lake Michigan, and slices of beach connected to one another, their backs to the urban sprawl, their faces out to the oceanlike horizon. The beaches were ecotone—a liminal space between the wild and the tame—full of animals. Their beauty and the depth and power of the lake were as shocking as if you dropped the Grand Canyon down inside Los Angeles.

Dex and I looked for animal bones in the low, wind-raked dunes of the beach, foxes beside the buildings watching us from the tall grasses at the edge of the sand. Dex poked at the body of a rotting salmon with a stick as we stood ankle-deep in drying lakeweed, and the live bodies of small flies, zebra mussels on a dead tree trunk next to us.

I'd read the articles my mother sent me about collapse, warming oceans, the Sound, the Gulf, the lake, the rivers of my childhood failed and broken by industry. I could picture her, the way her body had slowed as she'd gotten sick, left her work, no idea what was causing the fatigue and pain. I lay awake terrified by the statistics in the articles, the rate at which the world was warming, the exponentially increasing chemical threats to the water, the speed of glacial melt. The doctors told me when I got pregnant that Dex would have a greater than 50 percent chance of having PKD, closer to 80 percent. I didn't want to know. There was so much to be afraid of.

"Mom, they are rattles." Dex pointed to the mussel shells, his voice too young for the full sentences, tiny professor, little naturalist, playful, solemn—my dead father resurrected in miniature. He ran his fingers over the nest of zebra mussels in the damp sand. The shells clicked together in a light chatter.

Two weeks before Dex and I set foot on Monhegan together for the first time, a derecho moved over Chicago. Jonah and Dex and I had been at the beach on a hot blue day, but the storm pushed a wall of black clouds at the lake, and Jonah lost his wedding ring, and the sand emptied of people, and the wind ripped a tree from the ground next to us as we ran with the stroller. It fell three feet from us. If we'd been slow, it might have killed us. At home in the apartment, the house was hot, still, electric-charged, the storm over the lake on its way to Michigan. In the kitchen, Jonah pounded the kitchen cabinets, screaming about money. Dex and I hid in the bathroom, the solid wood door rattled every time. I closed my eyes, and I was nine kneeling next to my father, my mother locked inside the bathroom threatening to kill herself. I opened them, and I was thirty-seven, child in my arms. Sweat ran down the backs of my knees.

"It'll be OK," I whispered. I put my hands over Dex's ears. He pulled them off and looked at me, serious and quiet. I handed him mildewed bath toys from the box by my bare feet—a whale, a squid.

"Is it locked?" Dex whispered, squid in one hand, whale in the other.

"Yes," I whispered back. I picked leaves from his hair. I pictured waves.

"Fuck you!" came Jonah's voice from the hallway outside the bathroom door. The front door to the apartment slammed shut behind him, his feet thumping down the stairs and out to the street.

My breathing was so shallow I was dizzy. I unlocked the door and called my friend, and she came and helped me pack my bags and put Dex in a car seat. She waved to us as we drove away. From a hotel in Indiana, I called to tell Jonah I'd left him. The next day, I drove twelve hours alone from the derecho and the bathroom. Flat land rolled through sun and rain. I made sandwiches on the hot metal of the closed trunk of the car. I didn't rest, not even when I saw my mother on her porch in the darkness at the end of that day. She had cut her hair short, and when she walked from the front door to the steps to meet us, she limped. She was going gray more slowly than I was, and the lines of her face were still strong and beautiful; the young mother she had been on the Sound showed in her cheekbones and glacially pale blue eyes.

I told her that the line of wind had come from Omaha, across Chicago, that we followed it, news reports of destruction on every truck-stop television on I-90.

"Of course," she said, "it's all coming apart, the whole ecosystem."

In the morning, Dex found a dead mouse in her garden, held it up to me by the tail. "Mom, if we bury it in a box the bugs will eat it." His body vibrated with the future.

"He's just like you," she said. "Remember the cemetery you made for mice at the Great Barrington house?"

We buried it in a shoebox, covered the grave with smooth gray stones we found in her forest of laurel and maple.

"Next year," he said, his tiny voice a high lisp, "you'll see. Bones."

That night, I lay next to him on the futon in her living room, the woods flat black outside the windows; on the wall, a photograph of me at fourteen that my father had taken, sitting on a rock on the Maine coast that first summer I saw Monhegan.

My mother snored in her room. No one could diagnose her illnesses that moved in her as if the toxic horrors of the rivers she'd been cleaning had entered under her skin and poisoned her too. I had run home for safety, but I remembered that she was stronger when I cared for her, not when I needed care. Already, claustrophobia made my skin itch. Dex's breath was sweet. His cast was scratchy against my bare arm. The picture glowed on the wall. A cascade of understanding fell down on me in the dark. I had to get us to Maine. Something was waiting for us there.

A week later, Dex and I were on the ferry across Muscongus Bay in the Gulf of Maine to Monhegan.

The Gulf is a semi-enclosed sea that is part of the Atlantic Ocean. It is what's known as a shallow-water benthic zone, home to more than three thousand species of marine creatures and plants. Of those, there are 652 species of fish moving through the ecosystem, eighty-seven of which live there exclusively, thirty-two species of marine mammals, high baleen whale biomass in comparison to other marine ecosystems, more than 184 species of birds, 271 species of macrophytes—saltwater algae that you can see without a microscope—and more than 1,400 species of invertebrates. Most of these species are dependent on the temperate to cool waters of the Gulf.

The Gulf forms a roughly rectangular geological depression in the Atlantic Ocean, defined by its thirty-six thousand square miles of hydrogeologic terrain, the drainage site for more than sixty rivers. Warming water and rising sea levels happen in location-dependent ways, and the geology of the Gulf creates its fate. This happens in other places too. The Arctic is hotter relative to its own normal than to New England. Parts of Florida experience sea level rise more quickly than parts of the United Kingdom. Consequences are personal. The C-shaped depression of the Gulf once kept it cool; now it traps the heat instead.

When I was fourteen and it was 1985 and we took the ferry to Monhegan, the Gulf was cooler. My father was already on dialysis then, waiting for the kidney transplant he would get later that autumn. None of us had ever seen the island. We took a day trip from the family reunion of my mother's side of the strong, healthy descendants of Scottish immigrants on Southport Island, where Rachel Carson once lived. We sailed on the *Laura B* on a bright day after a storm, and the high swells made me seasick. I took to lying on the ferry bench, wondering why we'd come. When the mailboat landed, I shook the vertigo out of my limbs, stood at the harbor looking up at the hotel and the cottages, and I fell in love.

We hiked through the tall forests and sunlit swamplands to emerge from the claustrophobic single-track trail into the wide, bright salinity and wind on Black Head, the tallest cliffs in Maine. My father took my hand and walked me to the edge, where the basalt cliff face fell away, pine forest at our backs, White Head and Burnt Head marching away from us. Below my feet, I could see the surf crashing on rocks, and the seagull rookeries, and the ducks riding the waves at the edge of the land.

My mother, terrified of heights, clung to the tree line, far away from the drop. "Come back here now, Charlie! Eiren!" she yelled at us, her voice cracking, the edge of a sob in her. My father held my hand as we stepped as close as we could to the lip of the world.

"Just look at that," his face was set. I tipped my head up and looked out. There was nothing between me and the ocean.

Though semi-enclosed, the water of the Gulf depends on both the Gulf Stream and the Labrador Current. It also depends on the Gulf of Maine Coastal Current—a result of inland meltwater from seasonal snow entering the Gulf via its riverine ecosystem—something on the order of sixty billion cubic meters every year.

Before collapse, the Labrador Current controlled the pattern. It flowed directly into the Gulf, ensuring that the water along Maine's coast remained cold, its cold-water-dependent ecosystems sustained and hearty. As waters at the equator warm, animals run north—squids and lobsters and whales and plankton. In the Arctic this shift of water and creatures is called "Atlantification" and simply means that the Arctic Ocean is now, more so every day, the Atlantic.

On the *Laura B* that first time Dex and I went to Monhegan together in 2008, Dex sat upright in the swells, his body leaning into them, exhilaration on his face. Next to him, a woman cradled a small, white dog in her lap as it vomited into a bag she held to its mouth.

I tried to settle my stomach by looking at the line where the horizon met the sea. Monhegan emerged from the Gulf as a hill of green forest and gray cliff, a whale back itself like those described in old bestiaries. A whale made into an island, ready to pull the unwary traveler into the profound.

Dex and I hiked on muddy paths and slept while the sound of the sea came in through casement windows of our hotel without screens. I put it all on credit cards. I imagined repair. I imagined escape.

"I'll be better," Jonah told me over spotty phone reception.

Skies would gray and lower, lace curtains pushing back in the wind. Elation sang in my blood with the church bells in the morning. There was an averageness to the miracles on the island, a regularity of mist in the pine forests, tourist voices on the cliffs. I was pulled into it, a place I'd seen only twice after the trip with my parents, once with Jonah, once with my friend Robin on a day trip. We were sheltered and given over. I could feel Dex unfurl there. The rules of city childhood disappeared. He walked through mud pools on the trail in bare feet. He met strangers. He took his shirt off at the tide pools and lay on the long, coarse gabbro and basalt rocks and closed his eyes in the sun. After a few days, I no longer wanted to save anything at home. I wanted to save myself.

At Lobster Cove—my back to the rusting 1948 shipwreck of the *D. T. Sheridan*, Dex held up a snail, his face alive with wonder. He worked hard, hair in his eyes, to reveal the miracles to strangers, taking other tourists to the clefts between the rocks, to the silent places where the mussels and the barnacles grow.

"See," he'd tell them, surprising in his direct gaze, too little to be refused and too little to be believed, "there are the barnacles." I didn't know that his preternatural early language skills and focus, his obsession with animals, his vocabulary meant that his mind worked in unusual mysteries. That was the kind of child I had been. I didn't know my own mind did either. I approached him as I had wanted to be parented myself. I understood how to parent him because we were the same.

Every surface of the intertidal zone in the Gulf of Maine is covered in barnacles. Our feet caught them as we walked; they lay alongside periwinkles, covered the shells of lobsters. They are the only arthropods in the ecosystem that anchor to a single place for their entire adult lifespans, rooted with cement, feeding on phytoplankton, forming their shells in taller and taller layers as they age, more accumulation of information and history, the agglomeration of place.

They come together for protection, rely on balance in temperature and water quality and food resources. They seem endless and solid. They are instead a function of stability, not a promise of it.

A barnacle opens its shell to eat, and from it extend what look like fronds but are really legs—cirri—thin, covered in wisps resembling hair or feathers. Cirri sweep the water, carefully, deliberately, filtering the plankton from it, the barnacle dancing in place.

Years after that first year, when Dex was seven, a woman he'd taken to the tide pools, done with his eager manner, tried to stand up too soon from a barnacle observation. "Wait," he told her, "if you watch closely, you can see them eating." When the woman saw the barnacle eat, she looked at Dex with an open wonder in her eyes, caught from him, from the tiny grace of seeing a familiar creature in a way she had never seen it before. Dex looked at her. "Did you know that the ocean is acidifying?" he asked.

She blinked, the wonder disrupted, and shook her head.

"I wish I knew what the barnacles will do," he said, then led her off beyond my hearing to see more beauty.

Barnacles, like the rest of us, will have to change if they are to survive. They may, however, hold the keys to understanding how genetic adaptations prepare species for the coming changes to the oceans. Salinity changes in ocean ecosystems like the Gulf as the balance of salt water and fresh water is changed through melt, through altered currents. Barnacles are residents of the intertidal zone, a part of the liminal space we favor, Dex and I—the littoral zone. The barnacle, like us, loves the in-between. They have evolved to adjust the enzymes in their bodies to alter the balance of glycolic proteins that help them digest sugars from their food.

If they are in fresher water, they adjust the amount they glean. If they are drier or in more saline, they adjust back. Barnacles may also alter their osmoregulation patterns—the systems that balance salinity and water pressure in a body—in response to increasingly violent alterations in their ecosystems. They are learning to hold shifts, extremes, the either/or of the new world in their bodies.

Scientists are using the intertidal zone as a laboratory to understand how species survive and prosper in fluctuating extremes. The same methods used to show how barnacle genes adapt to those extremes could be used by entomologists who want to know how honeybees adapt to climate change, or by crop scientists concerned about productivity in stressful environments. Evolutionary approaches may be key in adaptation to collapse.

Monhegan was my adaptation, rebalancing my body. I would wake with a start, remembering the bathroom door locked against Jonah. I would put those thoughts to the side and walk out with Dex into the wilderness all day. Lying on the rocks with Dex at my side, the sun going down over the ocean off Lobster Cove, I could feel a shift. The warm rocks held me, geologist's daughter, as she hadn't. The science she taught me made them real. The sound of the waves overtopping the tide pools was a mantra that took me out beyond the body I feared.

Precarity made my marriage look like a good idea. The Midwestern nice of Dex's father a comforting opposite to my parents' bohemian chaos and inability to manage crisis. I ignored how his rage was like my mother's, how, like her, he'd offer care, then rescind it with criticism, that in the end he preferred it when I didn't talk about PKD.

I stretched my arms out to the side, the warmth of the stones on the bare skin of my shoulders, my wrists. My hand touched Dex's hair, flared out on the rocks next to me. I turned to look at

him, peaceful, eyes open to the sky. If his father couldn't acknowledge my illness, he'd never be able to acknowledge Dex's, if PKD
came for him. I knew then that there was no going back.

The island's restaurants would close soon. I took Dex's bucket
and emptied the creatures from it, filled it up with his scattered
toys, tied it to my backpack, and took his hand. We passed a couple
who waved. Dex hid his cast from them.

"Kiddo? Do you feel shy about your arm?"

Silence.

"Because you didn't do anything wrong. Daddy and I weren't
fighting about your arm. Your arm was just an accident. Daddy
and I are fighting because of us."

Silence.

"Because Daddy and I are not able to talk to each other kindly
sometimes. Even though we love each other."

We climbed over a set of rocks in the path, using dwarf pines
as handholds. On the other side, Dex held his arms up to be carried. I shifted the weight of my pack and grabbed him, so he
balanced on my hip. "Daddy loves you, and I do. He wants to talk
later on the phone."

"No, thank you. I want to play at Swim Beach."

"OK. We'll do that after dinner."

I would tell Jonah later that reception was bad. I would let Dex
rest. I would let myself rest, too, rebalancing our salinity.

We crossed the inlet of the Cove. I set Dex down and then
hoisted him onto my shoulders; his weight distributed evenly over
my back then, his shorts scratched against my sunburned neck. I
looked over the way we'd come. Lupines and young goldenrod
were covering the slope of a transitional meadow, glowing. I felt
how tired I was. I replayed how much money I had in my pocket,
enough for us to split a pizza we could eat on the beach as the sun

went down. I stumbled on a rock, caught us both just before I planted us face-first on the rocks, and thought, *This is how it will be from now on, his weight on my back alone.*

Maybe my father had thought of this, what it might be like to carry me himself. Maybe my mother had. They never threatened divorce when they fought, but quietly, once I was out of college, my mother rented an apartment, hired a lawyer, booked the movers, and left my father wondering how he'd missed her plans. He brought gifts to her new house and stood outside the door. She sent him away. He told me, "Kiddo, tell your mother I love her. Tell her I want her back." She never softened. He went on alone, and I took up the weight of his illness for her, and mine too. Now I was about to let go of the idea that Jonah had been carrying anything at all. I was about to take it all on. I thought, *None of them could make it. But I will.*

For a few months, though, I did go back. We tried therapy and medicine and bankruptcy and separation and work. Nothing changed.

"Daddy has *Alligatoritis*," Dex told me at night in the house after his father finally moved out and I changed the locks.

"What's that?"

"When an alligator gets a wildebeest, it can't open its jaws until it is dead. When it is dead, it can let go."

"Oh."

"Daddy can't let go."

"Sorry, Bean," I said that day on the island, "didn't mean to stumble," and I took a break, breathing, adjusting his weight on my back until I was ready to go.

"Will the beach be empty?"

"Probably. Unless people come to watch the sunset."

"I hope so. I don't want to talk."

I walked into the shade of young trees, blackcap brambles around me. He reached up with his good hand and touched the leaves, took a blackcap and popped it into his mouth. "Is that because you are shy about your arm?"

I felt what must have been a nod.

"Why are you shy?"

"Because," he said through noisy chewing, "when they ask, I have to talk about my body when I just want to talk about fish."

In the Coast Guard cutter, I was awake all the way to Damariscotta, if that fading in and out was wakefulness. Every wave felt more and more like a swallowing whale, like a whirlpool in the sea, like Scylla and Charybdis. The years on Monhegan were with me, the body I was trying to escape. Underneath me the wild Gulf of Maine, bigger and stranger and more complex than the Sound. Underneath me were all the creatures I loved and the water I came to Monhegan to visit, to write about, to study. Love of the place brought me, but so did fear for the place and for the creatures living there. In the wild weather, the warming water, were non-human kin who were as vulnerable as I was. I couldn't think of them clearly or of the web of consequences of collapse that I was working to untangle in my own understanding.

All I could think about in the boat—the thrumming of the engine louder than thought, my erratic breath and heartbeat mirroring the sound of the wind and the storm—was what might be there in the sea with us. Like the flower drawing of Dex's dream, my half-conscious mind fed me images of animals, seals, puffins, tiny sea butterflies, larval ladyslippers, lobster, branching zooplankton, right whales, salmon, and ocean sunfish. They were all accompanying me on the other side of the skin of the boat. I remembered Dex's words, *I have to talk about my body when I just want to talk about fish.*

Copepod

(*Calanus finmarchicus*)

The cutter landed at a concrete dock, and the low clouds reflected back the electric lamps from town, a novelty after the dark of the storm on Monhegan. They made it look close to dawn.

I had the particular instability in my legs that comes as you move from a boat to the land—the lightness in them, as if you are about to fly off the earth, then the heavy landing as they remember the gravity of soil and rock, and root you, willing or no, back into a body made of meat, not water.

I remembered in my bones that transition from a hundred landings—ferries, my grandmother's catamaran, the Boston whaler I took out fishing, the sailboat belonging to Ben's parents where I first saw bioluminescence.

In a nor'easter, you can't see bioluminescence. The water under the boat we were leaving was so roiled and turned by the storm, and we were all waiting it out, the humans, the animals.

The Gulf of Maine has a unique aquatic habitat, a water column in which distinct environments rise from the dark benthic depths at the very bottom of the sea to the surface, providing a nursery for many species, providing zooplankton for pelagic fish like herring and mackerel, which attracts storm petrels, shearwaters, predatory fish, fish-eating mammals, jellyfish that attract leatherback turtles. The Atlantic Meridional Overturning Circulation (AMOC), that infinity symbol of circulating water that stretches from the north to the south poles—mixing the water of the Atlantic, cold to warm, deep to shallow, and transferring life to the sea—makes the water column possible. But, as the Atlantic warms, the engines of temperature and salinity change. The more ice melts into the world's oceans, the more imbalanced they become, threatening to stall the AMOC and cut off both the swift currents that link one continent to the other, and also the warm temperatures that have kept northern Europe temperate for thousands of years. Some data report that the slowdown is increasing, and that there is a chance the AMOC will stop altogether, with catastrophic consequences.

In the Gulf, distant changes have local impacts based on geology. Greenlandic glaciers melt, the new fresh water enters the ocean north of the Labrador Current, and it is lighter than the saline water it meets. Fresh water stays at the ocean's surface, while the saline water sinks. This—so opposite from the actions of a normal ecosystem—results in a slowing of the cyclical and circular flows of the AMOC. The Labrador Current stalls. With the Labrador sluggish, the warmer waters of the Gulf Stream rush into the void and fill the Gulf of Maine with warmer water. The C-shape of the Gulf holds that water in its semi-enclosure, and the raised undersea plateau of the Georges Bank acts as a blockade to the water leaving of its own accord. The result is trapped heat and the fastest-warming marine ecosystem on the planet.

The dock in Damariscotta was steady, solid, and the ambulance was waiting to take us to the hospital in town. The EMTs moved us together, tucking Dex next to me on the stretcher, shielding our faces from the rain. They let Dex snuggle into me, still asleep or pretending to be so, and loaded the stretcher into the ambulance, while the rain pounded the ambulance roof, while the Coast Guardsmen talked to the EMTs, soaking wet, the waves piling up on the dock.

"Female, possible seizure. She's got a kid with her," said the first Coast Guardsman.

"A kid?" said the EMT.

"Yeah, they were alone on the island. We were told not to separate them."

Seizures are not part of the disease profile of PKD. When Jen asked if I had a seizure disorder, it had struck like a badly played piano key. *We don't have seizures,* I thought, *we have aneurysms.*

In the age of collapse, the rivers that feed the Gulf of Maine are making more runoff, because in this new era, there is more precipitation, and more rain means more erosion, more nutrients leaving fields and riverbanks. In the Gulf of Maine, this has meant that the water has begun to turn color.

The founder of the modern science of oceanography, Henry Bryant Bigelow, made a study of ocean colors. In 1912 he charted the colors of the Gulf of Maine in his schooner *The Grampus.* When compared to Bigelow's charts, the current Gulf is turning yellow. Aesthetically this is remarkable, but on a practical level, the darker color means less sunlight for the photosynthesis needs of plankton. A NASA-funded, eighteen-year study of the waters of the Gulf show that this loss of photosynthesis opportunities means a loss of phytoplankton productivity, which changes the productivity of the entire food chain. Zooplankton has been on earth for 540 million years; bioluminescence is the least of what it does.

All of the processes of the Gulf have to do with the building blocks of its life—currents and plankton. The Gulf Stream, which takes its name not from the Gulf of Maine, but from the Gulf of

Mexico, is a warm highway pushing north into the cold Atlantic. It has always skirted the edge of the Gulf of Maine, delivering animals to, but not across, its gates.

Ponce de Leon, the first European to document the Gulf Stream, wrote in 1513 that it was "a current such that, although they had great wind, they could not proceed forward, but backward and it seems that they were proceeding well; at the end of it was known that the current was more powerful than the wind." Sailors, especially early American whalers, knew of the Stream, rode it from the Americas back to England. But it took Benjamin Franklin to finally document the current on a map. His cousin, Timothy Folger, a whaler who had ridden the current on whaling ships in the North Atlantic, told Franklin about the secret shortcut whalers used to cross the Atlantic, the current that made faster travel possible. It wasn't new information to people who worked in the water, but Franklin documented it first and gets credit for that initial map.

The current has made possible the migration of people and the invasive migrations of countless species. There are ways to use it to track the health of the ocean, and its ripple effect on the health of animals. It moves plankton around in its grip, changing the feeding patterns of other animals, the nature of the ecosystem. It is what allows for the green and pleasant fields of England and the fisheries that feed the coastal United States. It brought with it both the chaos of the nor'easter and the waves plied by the boat that brought us to help.

They turned the ambulance around on the dock and drove toward the hospital; the pull of it brought back a wave of vertigo.

"Mom?" Dex asked. "Are we in an ambulance?"

"Yeah, kiddo."

"Are we off the boat?"

"Yeah, we left the boat."

"I liked being in the boat. It was like *The Nautilus*."

"We're going to a hospital so they can make sure I'm OK."

"Are you OK?"

"I think so. They just need to be sure."

"Will we go back to Monhegan?"

"I think so. That's where all our underwear is."

"OK." I could feel Dex will himself back to sleep.

The EMTs didn't use the siren, but I felt the speed of the travel. The vertigo returned, and the energy I'd used to force normalcy into my voice drained from me. Exhausted, I concentrated on slowing my breathing.

The storm had stirred up the smell of the sea. Phytoplankton produce the indelible smell of the sea in their dying, releasing a chemical they've used to survive in the ocean. Bacteria consume the dying plankton and convert that chemical into another, the resulting emissions providing the strong smell of the sea, which is used by seabirds to target plankton-rich—and therefore planktivorous-fish-rich—waters. Everywhere the smell of released death.

Copepods, called *Calanus finmarchicus*, were in that stirred-up sea. A mesozooplankton and perhaps the most important plankton in the Atlantic, they begin as eggs, maturing into eyeless, shrimplike adults—no more than two to four millimeters in length—miniature crustaceans with what look like claws, legs. They are found all through the northern oceans. Their eggs are the primary food for central species in the food web—cod, alewives, menhaden, herring—providing up to 90 percent of the diet of larval red fish in the Gulf of Saint Lawrence. They are the most significant copepod in the Sound's environment. They eat the *A. fundyense* that cause red tides, and when they do, there is a substantial drop in *C. finmarchicus* egg production, a threat to future generations. In the Gulf as in the Sound, the interconnections of creatures echo over time.

In the ambulance, interconnections echoed too. We pulled up to the Damariscotta hospital. They moved our stretcher into the blinding light of the triage room, the storm receding behind the closed doors. There were so many days of walking my father into over-lit hallways—after infections or strokes or transplant rejection—a hundred bleached corridors, washed-out lights, the smell of disinfectant, the weirdness of the middle of the night, when you are no longer sure of any story, and all you can do is ask for help, inarticulate, *please, here he is, please figure it out.* Now the child in tow was mine.

From the bright hallway, we were deposited in a room with its lights off. The nurse looked down at Dex, still, unaccountably sleeping. "Let's keep it quiet in here, OK? Get you some rest. It's late." The volume of her voice was a half measure in the dim room, neither a hush nor normal.

"OK," I whispered.

"He's so sweet." Her voice dropped to match my volume. "You hungry?"

"No."

"Well, I'll bring you something anyway. Kitchen's closed, but I can get some crackers in case he wakes up," she whispered.

She left us, shutting the door behind her. There was a huge light over the bed, turned off, like ones from a movie operating room. I thought I could smell blood, or salt. The seizure had slowed time. The feel of the wet twilight air at the crossroads by the apple tree on the island permeated my skin, the memory of my head shifting on the gravel of the road as I tried to concentrate, flashes of Dex on the cutter with his hair plastered back by sweat from the heat of the boat's interior, the dry suit bagging around his limbs.

The PKD kidney, growing and changing and obscured by cysts, operates for a long time as a locked black box, with secrets and mysteries it will not willingly reveal. I had no idea whether mine were failing already. I could not remember if seizures came with aneurysms or if stroke produced seizures. I feared that whatever had happened was an early symptom of loss of kidney function. I tried to remember the symptoms of a stroke.

Hospital time crept in. Lying on the triage bed, through the portal window in the door, I could see the hallway's overheated, white fluorescent glow.

"The doc wants to start you on anti-seizure meds, just in case," the nurse said when she came back with crackers.

"Are they safe for kidneys?" I asked.

"I don't know."

"I can't take them if they could damage my kidneys."

"You'd refuse them."

I nodded.

"OK. I'll check with the attending."

It is too much for zooplankton, as endothermic creatures, to survive marine heat waves. In the face of them, zooplankton move north to adjust. They don't move like the rest of the ecosystem, since they can only drift with the currents; as the currents stall, they move by attrition. In their usual haunts, they die out or reproduce less. In the regions that are less heated, they begin to produce more. In this way, their population shifts, permanently changing the Gulf of Maine.

"The attending doesn't know, but he thinks it would be bad for the kidneys. But if you have another seizure . . ."

If I took the drugs, I might stave off another seizure. But if the seizure had happened for some other reason, the drugs would mask the problem, and I'd harm my kidneys in the process.

"I don't need the drugs," I said.

Her eyebrow went up. "OK?"

"Can we just wait and see? Can we wait and see if I'm better tomorrow, if I can get back to the island?"

"Alone?"

I nodded.

"Honey, I don't think that's happening."

American Lobster

(Homerus americanus)

They moved us from the ER room to a hospital room to "keep you for observation and see if you have another seizure," the nurse said. Rain battered the glass, obscuring the colors of the land and water. Dex lay next to me while I passed in and out of sleep. I went through the bag that Jen had the young man pack for us on the island—clean underwear, Dex's plush octopus, a book, a sweater, a coat, and a pair of size ten boy's pants, but no phone charger.

A new nurse came in to check my vitals; she tipped her chin toward the window. "They say it's a nor'easter. Power went out at Logan." As if to illustrate her report, the wind picked up and distant thunder rumbled. The daytime sky was dark. "You can't order breakfast anymore," she said, handing me a menu, "but I can get you both a tray. We're not supposed to, but I bet he's hungry."

"I think he will be."

She saw I was uneasy. "Take your time," she said, smiling, "no rush." She handed me the television remote and put an extra blanket gently over Dex.

"Looks like mine did at that age." She almost reached a hand toward him. "Hope he gets some sleep."

"Did the results come back from the CT scan?"

"I think so. The doc will be in to talk with you about it."

"No aneurysm, though?"

"Honey, if you had one of those, there'd be lots more drama."

I read the menu. There were lobster rolls for lunch.

In the divorce, in the whirlwind of parenting, I allowed the focus to slip off PKD and onto everything else. Dex came first. In session one day, our mediator asked Dex's dad: "Jonah, aren't you worried that the stress of this ongoing conflict will hurt Eiren's health condition? Her blood pressure? By setting up conflict and financial stress, will she have problems that will impact your kid?"

Unironically, he responded, "What health condition?"

In researching denial, as I did on nights alone in the condo we had shared, the one in foreclosure, I read about anosognosia. Anosognosia is a clinical condition, one that occurs in severe mental illness like schizophrenia, memory disorders like Alzheimer's, or in brain injury such as a stroke. A portion of the patient's brain simply refuses to believe that the patient has any disease at all. It isn't a choice, it is a blind spot, a crucial piece of information completely missing from the person who needs it most.

We often say that people are in denial about ecocollapse, as if there is a collective waving off of information. But we have entirely lost the part of ourselves that recognizes ecocollapse. When a patient has anosognosia, the impacts of the condition on their treatment are profound. They'll resist medicine, therapy, even discussion of the problem; planning to cure or to cope is off the table. In ecosystemic collapse the effect is the same. I had no context for seizures. I viewed everything that happened to my body through what I knew about PKD.

Anosognosia can be a comfort. I'd wake up from a deep sleep in my rented college room when I was twenty-three, the year after I learned I had PKD, and fail to remember my diagnosis, anosognosia my drug. In the years since my divorce, I went from a woman with a partner, an apartment I owned, health insurance, steady work, to someone squatting inside a foreclosure, falling ill, lacking money to care for herself, with primary responsibility for a child vulnerable in a neurotypical world. By every measure of adult life, I was ruined. The low level of anosognosia I had for years was necessary for daily life.

What we are using to cope with the specter of collapse is more like anosognosia than denial. We expect that the big things—tides, animals, currents, seasons, livelihoods, home—will remain over time, regularity that can be transferred between generations. Even difficult things feel as if they will go on—winter lobster fishing, all freezing waves and grueling work, is a part of what we expect to bequeath in families dedicated to the work. Increasingly, we can't. Managing the emotional response to that fact remains a problem of scale. Shifts like the ones we are facing in collapse are simply too enormous, too permanent, too gutting to be held in the scales of time and space that humans regularly encounter. We do not know how to argue with anosognosia, with a flat refusal to look at reality, in ourselves or in others.

By 2025, the population of Maine is predicted to reach 11.4 million people, with a little more than half of them dwelling along the coast. In Maine alone, the marine-connected economic sector, even ten years ago in 2014, provided more than thirty thousand jobs. In a 2018 *Reuters* special report, Maurice Tamman wrote: "Until this century, only about 50 percent of all fishing revenue in Maine came from lobstering, according to U.S. fisheries data. In the 2000s, that started to steadily rise until, in 2016, it topped 82 percent."

Maine lobstermen catch more than 80 percent of America's entire lobster haul. They set a record of more than 132 million pounds of the crustaceans in 2016, worth more than $530 million at the docks. But in 2017, the catch began to decline, and now it fluctuates year to year and hasn't retaken that lofty height. Proportionally, the lobster catch continues to be the biggest part of the Maine fishing industry. According to data collected by the state's Department of Marine Resources, in 2022 the fishery earned more than $574 million, of which more than $388 million came from lobstering.

In 1905, when the State of Maine first began to regularly document water temperature as part of its lobster landing data, the temperature buoy in Boothbay Harbor, mid-coast on the Gulf of Maine, registered a water temperature of 7.8 degrees Celsius and a landing worth $1.4 million in lobsters and totaling more than 11 million pounds. In 2022, those numbers were almost 98 million pounds of lobster, worth almost $400 million, and the temperature was 10.9 degrees Celsius.

When the doctor arrived, he was young, kind. Dex watching endless cartoons on a loop on the television suspended above us, *Steven Universe*, his eyes never wavering from the screen. The doctor spoke over the sound of the show: "It just looks like a seizure. But you don't have a history of seizure disorder?"

"No. Just a history of PKD."

"Your blood work is OK, no kidney failure, but your iron and vitamin D are low, you're dehydrated. Are you eating OK?"

"Not really. It has been a tough year."

"You need rest." He looked at Dex. "Is someone coming to help you?"

"No."

"Well, you should get someone here."

"The nurse is trying to find me a phone charger so I can call someone. My car is in New Harbor."

"Oh, you can't drive until we know why you had a seizure."

"Really?"

"Yeah. Seizures can be caused from lots of things. Your kidneys aren't failing. You didn't have a stroke or an aneurysm, No evidence of brain damage or heart issues or anything other than extreme exhaustion. But, why so extreme? You'll be barred from driving until your home doc runs another battery of tests to completely rule out a seizure disorder or something else. And you'll want to check in with your home nephrologist. They'll need time to make sure. We can discharge you, but only your doctors in Chicago can clear you for regular work and for driving."

"I drove here from Chicago."

"It's standard with a seizure. No driving for six months."

"That would mean I can't drive back home."

"All the better," he said. "You shouldn't be driving those distances right now. Whatever is happening, your body isn't up to it."

My mother was closer than Chicago, six hours away in Massachusetts, but she couldn't drive at night, or for more than an hour at a time. She was barely competent driving stick; my car was stick. No matter what, my car had to get out of Maine. Jonah still didn't know why I hadn't called him the night of the accident. My phone battery was dead, and I couldn't even see the text messages he must have been sending. Somewhere in my fog I realized I was risking a lot more than my health by failing to call him from the hospital room. If the staff thought I had no help, they might call family services, track Jonah down in Chicago, have social workers take Dex into temporary protection. It was a miracle he was still sitting next to me.

A Rolodex of people I'd never leaned on ran through my head. I'd held friends at bay and made a show of competence. I'd been so happy that those relationships existed at all that I'd never tested them, never opened up to the vulnerabilities at the core of my life. I had made myself isolated, stuck, frozen. Wonder alone wasn't going to get me what I needed. Anosognosia wasn't going to make the problem go away, it was coming like a wall of wave.

"Look," he said, "I know it's hard."

"It's hard."

"But there could be lots of simple reasons for your seizure. If you don't have another one while you're here, we can discharge you. And your doctor at home will figure it out."

"OK."

"OK, we'll get you out of here when it's safe. Back to the island."

"Right, back to the island."

Monhegan's harbor has a flesh of tourist economy stretched over the bones of a fishing town. They follow protocols that the lobstermen of Maine voluntarily established to protect the species in the face of declining fisheries in southern New England. They notch the tails of breeding females and send them back into the water; they release the smaller animals, making sure they live to grow and breed. This practice—and the protection afforded by the usually cold waters of the Gulf—has produced something of a lobster boom in the first decades of the twenty-first century. As the lobster fisheries of New York and southern New England have failed, the lobsters have moved north. If Maine's Gulf warms enough, the boom will flow to Nova Scotia. There is a three-mile radius around Monhegan, the shape of a fisherman's hat or an antique diving helmet, that cannot be fished by anyone who does not live on the island and hold a commercial license linked to the island's protected fishery. Those licenses are only held by a few people who live year-round on Monhegan, which legally cannot exceed seventeen, with room for more people to come and work on Monhegan. The only catch to that availability is the scarcity of affordable housing for working families. Lobstering on Monhegan takes place during the winter when lobsters build back their hard shells and move offshore and lobstermen need a deep-sea license to bring in that catch. Boats can run anywhere from $28,000 for a used vessel to $350,000 new built, traps average between $150 and $200 each, and the average lobsterman runs four-hundred-plus traps on their own. Monhegan lobstermen may not exceed four hundred traps. Ninety years ago, the lobstermen of Monhegan put a voluntary cap on the size of lobsters they'd bring in. On Trap Day, all the fishing families begin that season together, a nautical barn raising.

When you understand vulnerability and you resist anosognosia, you can create sustainable structures built on community that protect the people inside them. I assumed that my marriage to Jonah would do that. But in divorce, that community was revealed to be empty, broken, antagonistic even, leaving me to flounder and to find community outside the marriage. Alone in the hospital room, I realized that this was the time to discover if that new community was sustainable and capable of handling crisis.

"Do you want to try lobster roll for lunch?" I asked Dex when the doctor left.

"No. No lobster." Dex went back to *Steven Universe*.

At five, Dex asked me to order lobster so that he could dissect it. At the picnic table at Fish House on Fish Beach, he sat at my elbow, helping me pull it apart, watching every messy gesture, the claws broken off and broken open, the meat extracted and buttered, the tail removed, the legs sucked dry, the tail meat separated in the wet mass of shell and parts and finally consumed. I saved the shell to throw to the seagulls. He was silent through it all, wary, his eyes narrowed in curiosity.

At six, he begged to choose the lobster so we could do it again. I was reading David Foster Wallace's *Consider the Lobster*, sitting with the book at the tide pools and paging through a description of the lobster's neurobiology, describing the pain they feel when cooked. I looked out at Dex playing in the water. He was deeply pragmatic about the reality of animal life. In a *Planet Earth* episode that he liked, there was a scene in a watering hole where wildebeests came to forage and drink. A crocodile, launching itself from the still water, grabbed the wildebeest by the throat and dragged it under. Sigourney Weaver intoned, "The crocodile will hold the creature, many times its weight, under the water by the neck until it drowns." *Alligatoritis.*

I found myself holding my own throat while I looked over at Dex. "Are you OK, sweetie? That was pretty violent. Are you worried? How do you feel?"

"Bad for the wildebeest. Good for the crocodile," he said.

We went to the Fish House early to pick his lobster. "Hello, I'm Dex and I'd like to order a lobster for dinner. Can I meet it?" he asked the woman behind the high counter.

Startled, she nodded, then took him to the big case, put a milk crate on the floor, and let him stand up on it and look inside. "I don't see the one." He looked at her, gravely.

She understood. "Of course, come with me." And they disappeared into a room at the back, where more coolers were hidden.

He came out grinning. "Got it." And we walked back to the hotel, to read and wait.

In the afternoon we walked to Lobster Cove. "Will the lobster hurt when they boil it?"

"Some people say they can feel it. I am not a lobster, so I'm not sure."

"And those lobsters were just out there?" He pointed to Muscongus Bay.

"Yeah, that's where the fisherman catches them." I didn't mention that the only lobsters at the Fish House were caught many miles away, outside of the Monhegan lobstering area.

"So, there are lobsters out there, right now?"

"Yeah, kiddo."

"And they're like the baby ones at the salt pond?"

"Some of them are old. And some are big, like the one you chose. There are every size and age out there. Even really huge ones, like that big albino we saw at Shaw's Wharf."

He was quiet again, and we scrambled over rocks to stand and look at the view back toward the mainland. We could see the pointed end of Manana, the sun getting lower. "I think we need to go there before they cook him."

"Your lobster?"

"Yes. I don't want to eat him. I want him to go back." Big, silent tears came down his cheeks. Standing in the magic-hour light, it was hard not to think he looked like a painting, *Boy with Regrets Considers the Lobster*.

"Can we put him back here? Will he get home?"

"Of course, kiddo," I said.

He locked his eyes on mine, an unusual gesture for him, eye contact generally painful and rare. "I didn't know. I didn't think about it. I didn't think about him." We made it back and paid for Dex's live lobster and put him in a bag. It was plausible we were taking him home to be cooked in a cottage. We carried the lobster through town—wriggling—until we reached the Cove. We waded in and sent him home and sat on the seaweed-covered rocks.

"Do you think he had a family?"

"Probably. He was big, he might have been able to breed."

"They'll be glad he's back."

"I bet they will."

"Do you think he'll tell them? About the boxes? About the traps? Do you think he'll teach them what happens? Help them so they don't get caught?"

"I bet he will. He'll look at the nice bait inside the traps and point to it and tell the kids, 'Don't you go in there!'"

"Maybe he'll tell them about me."

Like most aquatic creatures, lobsters have a golden mean temperature they prefer, cold enough to hibernate and procreate, warm enough to have long summer molting seasons. They also need waters cold and regular enough to produce their prey, including *C. finmarchicus*. Juvenile lobsters, referred to by researchers simply as *lobster babies*, are the best indicator of whether a lobstering region is healthy and will continue to produce a strong fishery.

In a study, University of Maine researcher Rick Wahle looked at the density of lobster baby populations in twenty-three sites along the Atlantic coast of North America, from New England to Atlantic Canada. In sites in the southern regions, the number of lobster babies observed was in decline. In sites in Nova Scotia and Atlantic Canada, those numbers saw an increase. This correlates with additional research from Joshua Carloni, a researcher with New Hampshire's Fish and Game Commission, who looked at data gathered by the Seabrook Nuclear Power Plant from waters surrounding its coastal location. The data showed that those waters were full of egg-laying lobsters, but without lobster babies. The water was also missing *C. finmarchicus*, the babies' high-fat food. The lobster babies were starving to death in the wake of the plankton's move north to colder water, and so the nurseries moved north with their food.

News about the nor'easter filtered in through the television, the nurses. There had been tornadoes in Connecticut; hours passed with me dizzy and unable to stand without shaking, phone numbers trapped in my dead phone. A new nurse came on shift; she saw Dex in the bed with me, bored and withdrawn and one eyebrow went up.

"We're on our own," I blurted out. She cocked her head. "My mom's coming," I lied. "She's on her way, so we just need to be here together until she comes." Fear rippled through me, fear the nurse might call child protective services, and Jonah would use the crisis as a way to take custody.

"OK, hon." She was curt. "Kiddo, do you want some applesauce?"

"OK." Dex's voice was flat.

"Thank you," I said, and the nurse left without saying anything.

Nor'easters are complex storms that sweep up from the sea between Georgia and New Jersey, following both the jet stream and the Gulf Stream northeast along the United States' Atlantic coast, combining warm ocean and cold air. They hit hardest in the cooler months and have caused both rainstorms and snowstorms carrying gale force winds. Those snowstorms have been some of the most epic and memorable in the Northeast, including the New England blizzard of February 1978, which shut down my kindergarten. My parents and I walked the streets of the mill town we called home then, landing at Mike's PX, the corner store manned by Mike, a World War II vet who fed me hot chocolate with whipped cream that he made himself.

One summer, a nor'easter hit Dogshit Beach, and I lay in the dark twin bed, sand in my sheets, the power going out, the sound of the wind on the uninsulated cottage walls, the rain pouring down the window glass. The storm was so loud it sounded like roaring, like a train, like a lion. It sounded like the sea was falling on us, like the rain would pour inside the walls, like the wind would come for the roof. If I think of the word *fear*, my mind will call up that night, that room with only me in the narrow bed, the darkness, the window washed as if by waves. It is deeper than anything, that fear, primal and wordless. But then I could hear someone in the next room open a door. I thought, *Please, please come find me, please be here with me.* And then the door to my room opened, and the storm got louder, and my mother was there, framed in the light filtering in from the bigger windows in the kitchen. I started crying, relief washing over me. She came and sat on the edge of my bed, balanced next to me on the mattress.

"It's OK, Monkey," she whispered. "The storm is big, but it will pass."

"Stay with me," I said, crying.

"I promise. I'll be here until it passes."

And she sat petting my hair as the wind battered the roof until I fell asleep.

Warming oceans make for larger storms, and the whole weather ecosystem is ramped up. Unpredictable weather creates a phenomenon called nonlinearity, where patterns one used to track with some ease lurch between poles of heat and drought, storm and shine, and do not run to plan. But, in searching records for the date, there is no mention by the weather service of the storm I watched from the hospital as a nor'easter. The word *nor'easter* was in the mouths of nurses. Jen said it when she told me about the boat. It was on the television in the room. It was said by the Coast Guardsman, by the EMT in the ambulance. Somehow it didn't pass muster in the long run.

The danger in my marriage and divorce hadn't passed muster as abuse within an overburdened system of family courts. In a counseling room, I was shown pie charts listing all the kinds of abuse in relationships, emotional, financial, physical. "All of it is dangerous," the counselor told me. "You aren't held to a standard of abuse only counting if you've been punched." But the courts where we lived considered outbursts, emotional abuse, financial manipulation, threats to be problems that middle class families could deal with on their own. It wasn't stated law, but it was implied everywhere that I should keep my counselor's words to myself, Jonah was just angry. I was educated and strong enough that his feelings were my problem to deal with, no different from the work of thousands of women in the court system at the time. Illinois's vibrant fathers' rights movement had made divorce a hell for women married to abusers of many kinds. I knew friends who'd been put in the hospital by their husbands with collapsed lungs, thrown down the stairs. In court, the judges looked down at them and said: "You've got a lot to prove, little lady. He still wants to be a father, that's something," and drafted joint parenting plans that entangled mothers with abusive fathers for years with no escape.

I'd read articles about mothers with bad divorces who were diagnosed with cancer, the custody fights in the courts where fathers pulled children from their dying mothers' homes. I gave away everything just to protect Dex. I minimized PKD in public, because I worried that if Jonah understood, he'd take even more of the precious time I had with Dex. Perhaps he'd stop just shy of harming me that way. Perhaps he wouldn't. I didn't want to take the chance. I tried to scaffold a life with both households that would keep Dex balanced, Jonah calm. I paid for my own lawyer. I let my mother sell her house to help us. I let it all burn down to keep Dex safe, all the time terrified that my PKD would upend everything. I disappeared into anosognosia to survive, living like there was no disease at all. I had nightmares of Jonah throwing me off cliffs. Sometimes, those dreams ended in rescue by old boyfriends, by friends. Sometimes, no rescue came.

"Dex, do you want this applesauce?"

"No."

"Is there anything you like on that menu?"

"No."

"Would you try a grilled cheese?"

"OK."

A lyric from one of my songs came into my head, "the wicked-ness of wasted time." The wickedness of wasted time is anosognosia's main project, spurring collapse while the patient refuses treatment. In the Sound, it wasn't only the hypoxic water of 1987 that changed how the community responded to the death of the estuary. The lobster fishery that supported so many residents of the Sound began to decline with a massive lobster die-off in 1999 and was lost through a series of midseason closures beginning in 2013. The lobstermen of southern New England used to bring in a huge proportion of the catch in the United States. In the 1980s and 1990s, Maine only accounted for about 50 percent of the lobster harvest nationwide. According to the Tamman article: "Between 1960 and 2000, Connecticut and Rhode Island in southern New England accounted for about 15 percent of the lobster harvest. Since 2010, however, lobster catches have collapsed in both states, with a combined haul of less than 2 percent." By 2015, the Connecticut fishery that employed hundreds of lobstermen had dwindled to less than one hundred. Those who left looked for new work or hoped to find new ways of harvesting from the sea. As in Maine, many turned to farmable fish and mollusks, oysters. But the lobstering has not returned. It likely will never come back. Not as long as carbon continues to flood the atmosphere, not as long as we pay the price for all the carbon that has come before. Until the oceans cool again, lobsters will keep moving north.

"You need this tray, hon?" the nurse asked, looking up at what began to feel like one endless *Steven Universe* episode on the TV screen. "That show still on?"

"It is a marathon," Dex told her with the most energy I'd seen since the accident. Garnet was singing a lesson about knowing yourself.

"No, thanks. I don't need the tray."

"No one liked it?" she asked in a confidential whisper.

I shook my head.

"They might let you leave tomorrow. Your mother can come get you, right?"

"Right," I said. "My mother can come."

"Oh!" She put her hand in her pocket. "We found you a charger!"

She walked behind my line of sight, plugged it into the wall, handed me the cord, and I plugged in my phone and started to call all the people I'd lied to about how sick I was.

Brook Trout

(*Salvelinus fontinalis*)

Martha answered the phone, and Amy, and Molly, and Amanda. I spoke to all the women I'd chosen as sisters in the years I'd lived in Chicago. They all thought I was calling from the island. Then I told them where I was. No one could get there for a few days. Somewhere in the Midwest, though, they began to plan. Martha told me she was buying a plane ticket. The phone flooded back to life in messages from Jonah asking why we'd failed to call. I called him. I whispered to Dex, "Don't tell Daddy we're in the hospital. I'll tell him later." He nodded solemnly, then spoke to his father about the island and the trip and *Steven Universe*. It felt awful to lie. It felt like keeping Dex safe in tricky waters.

I called my mother last.

During my divorce, she'd lived in Chicago for a year. She moved to help with childcare, but in that time, she broke an elbow and sprained a wrist, had cataract surgery, had colds and the flu. I spent more time ferrying her to hospitals than she spent caring for Dex. When she said she was going home to Massachusetts, I was as grateful as I was gutted.

"Where are you?" she asked, sounding like she'd just woken.

"A hospital in Damariscotta. I had a seizure on the island. I need you to drive me and Dex and my car to Massachusetts.

Martha can fly in and drive us back to Chicago, but she can't get to Maine. You're the closest."

"And you can't drive? Where's your car?"

"Not for six months. Until they know why I had the seizure. The car's at the hotel near the ferry. We were supposed to pick it up when we came back from the island."

"So, I can't drive up there, or there will be too many cars?"

"Right."

"Right. And you have a stick shift? I don't know if anyone can drive me to Maine."

"Can you get a bus?"

"Let me see."

"OK. Our stuff is still on the island. We have to go back for it."

"Well! I could get a vacation out of this! Finally get back to Monhegan."

"OK, Mom. But I'm in the hospital, you know. This isn't a vacation anymore."

"Well, I guess so. But, you know, I haven't been to Monhegan since your father."

"I know."

The hospital in Damariscotta was like all the hospitals with Dad. I'd been in so many since my parents divorced. I was the one answering his calls, *Honey, I think I had a stroke. Honey, I have an infection. Honey, I lost the kidney,* deep purple bruises on his arms like tattoos from wrist to elbow, the outline of the fistula thrilling against his skin. Technology changed, and I didn't notice the speed. With his first transplant, he acquired a virus, cytomegalovirus (CMV), that in another two years would be routinely screened out of transplanted kidneys. He was in a drug trial to combat CMV in the body of transplant recipients. He was in the control group. He lost the kidney.

When his second kidney transplant took place, the CMV woke up in his body, his suppressed immune system allowing it to begin transforming. Two years after the transplant, he was diagnosed with an infection, three years after that with a form of CMV-induced herpes zoster. Two years later it transformed into Kaposi sarcoma, the skin cancer HIV and AIDS patients routinely suffered. Each form of CMV mutation was treated, each form went into remission.

In 2000, when he had the first symptoms, my father didn't call me to tell me. Duffy and Jack, his two best friends in Boston, did. Jack and my father and Tommy, who lived in Brooklyn, were friends from high school in New Jersey. When my father dropped out of school to take care of his family after his father's death from PKD, they stayed close. Jack and Tommy went on to college and graduate school, to professorships in engineering and city planning, while my father finished his GED and learned to be bohemian.

"Your dad, honey," Jack said over the landline, "has he called you?" Jack was tall and thin, with a patrician accent mixed with Jersey. He had silver hair and a permanent grin and a kindness that let you forget his wicked sense of humor, his fierce intelligence. He taught aerospace engineering at MIT, but was too modest to mention it. Duffy was shorter, silver haired, with intense blue eyes and a wryness to her smile. She radiated a tough Yankee version of empathy and care, the kind that felt earned once you got it.

"No. I left him a message a few days ago. I'm about to leave town."

"Oh?"

"Yeah, I'm going out West on a road trip. I just took a new job in an office."

"Your dad is in the hospital," Duffy said on the extension. "And he's not doing well."

"What's going on?"

"They don't know. Something in his abdomen? It keeps swelling up with fluid. We've been going to see him, you know, help him out. But he's going to need you. He's supposed to come back from the hospital in a few days, but he can't be alone. We're going to bring him home here, to the guest room, you remember, the yellow one you slept in with your mother when he had his first transplant? And he can stay here awhile, but you should come. It isn't dire, but you should come."

I was twenty-nine. I had just taken the first regular office job I'd had since college. Dex and Jonah were years away. I was about to try to shake off a breakup with a drive to the Grand Canyon alone, in an attempt to forget Lawrence, the man I'd loved since I moved to Chicago.

"Marry me," I'd blurted out one night, four years before my father's hospitalization, after we'd seen a show. We loved each other—I knew that—and I expected an embrace, a nod, a smile. Instead, he laughed a panicked laugh.

"I can't," he said.

I was surprised but not deterred. "You'll change your mind," I said confidently. But he never did.

I needed the trip. I'd been so broke, working freelance teaching jobs, that I hadn't taken a break in ages. I was afraid of traveling alone, afraid to be without Lawrence no matter where I was. Now, I was poised to find something of my old freedom, and here was the call.

"Yes. Of course," I told Duffy as my mind raced. I had a rental car booked, a week away from Chicago planned, then I was starting a new job. "I can get out there a week from next Saturday, for at least a few days, get him settled somewhere, find out what services there are to help."

"Good. Because they won't let us help him with big decisions—we're not family. I can give you his contact information, the room he's in. And you can stay with us when you come."

"I'll call when I have plane tickets," I promised. Then I called my father.

"It's nothing!" he said. "If you don't go on your trip, I won't forgive you."

I was confused, torn, but consented without examining my reasons too closely. "OK, but I'll see you in two weeks."

"I'll be fine by then!"

And so, I called my father from pay phones all over the West. "What's it like there?" he asked from the hospital.

"There's a forest fire all through Los Alamos, close to Santa Fe National Forest," I told him from a gas station parking lot, smoke billowing over the range.

The weekend I flew back, they still didn't understand what was happening. Tommy came up from Brooklyn, and we sat in the hospital room together. He was direct, funny, the second father I'd known my whole life, reticent with praise. Like my father, he called me a smartass. A Greek whose family all lived to one hundred or more, his compact, athletic body was strong and sure; he danced at weddings with younger women, drove too fast, never ceded an argument. He read the *Times* and watched tennis with my father on the tiny hospital TV. He never mentioned the illness. They might have been in Tommy's living room in Crown Heights with a bottle of retsina between them.

The doctor and a crowd of residents came to the door. "Ms. Caffall, can we speak to you?" I nodded and followed them into the hall.

"He's got lymphoma. Cancer. That's what's in the fluid. It keeps moving around his body cavities. We've drained so much from his abdomen. We'll have to keep doing that until it stops. I'll come back with the team and let him know what the next course of action is when we have it. He knows it's cancer. He gave me consent to tell you."

I let myself back into the room, where Tommy and my father were laughing. There was a feeling of gravity, collapse that had become familiar, dislocation, in time, in place, in my father's body as it changed and changed, and the ripples of those dislocations dislocated my life.

From then on, each time I started a job or finished a record or planned a tour, Dad got sick. After my second record was finished in the late nineties, I flew to Manchester, New Hampshire, to start an East Coast tour, my lap steel guitar in the overhead compartment. My father drove up from Boston, where he'd moved after the divorce, to meet me at the airport. He was my ride to my first gig. At the gate, we hugged. "Honey, I think I've had a stroke," he said.

"Jesus, Dad, we have to go to the hospital."

"I think I can still drive you to the show. I drove here."

"No, that's not happening. You and all the drivers on 128 are lucky you're not dead."

I canceled the gig. In the hospital, I sat with him for two days, canceling another show, and waited until the results came back from CAT scans and MRIs, his brain a dotted map of mini strokes, his voice slurred.

"I'll be OK, Jack and Duffy will come. If you don't go tomorrow, you'll miss playing CBGB. You have to do that show for me."

The bronchitis I'd caught at the hospital didn't emerge until the night of the show. Tommy came from Brooklyn. My mother drove down from Western Massachusetts in a blizzard. The two of them sat together in the room, nearly empty because of the bad weather, and beamed at me. My father always wanted me to tell him which band I opened for. He'd buy their records and ask his younger friends about the musicians. "My daughter opened for Vic Chesnutt; do you know him?" he'd asked the barista at the sculpture museum where he'd gotten work as a security guard once the strokes made his hands too shaky for carpentry.

"Yeah! He's cool!"

On the phone with me he'd say, "Rob at work says Vic Chesnutt is cool."

"He is, he's great. You'd love his music."

"What record should I get?"

"Try *West of Rome* first. That's a good place to start."

Still the sickness surprised us both, each time.

"Are you going to get settled in at Jack and Duffy's house?" my father asked from the hospital bed, lymphoma flooding his body, his abdomen and, eventually, his lungs.

"Yeah, that was a long flight to get here."

"Did they give you a cool rental car at least?"

"You know, it is kind of a muscle car."

"And you know how to get to their house?"

"I do."

I drove to Arlington to the big, white house on the hill where Jack and Duffy lived when I was fourteen and visiting my father in the hospital for his first failed transplant, the one we didn't know would result in the lymphoma filling his body fifteen years later. Duffy poured me a glass of orange juice, and I took a box of Wheat Thins into the big library at the back of their house. Walls of shelves and books went up two stories, wreaths made of mussel shells and paintings Duffy had made of salt marshes lined the high walls. The French doors looked out at her darkened garden, and she steered me to the white couches.

"Do you have anyone to help you with this? That boyfriend?" she asked.

"Lawrence?" His name still hurt to say. "We broke up a few years ago."

"No one since?"

"No. He was special."

"You didn't want to break up?"

"No. I wanted to marry him."

"And have kids?"

"I don't know. I don't know if I should. Dad might not have had me if he'd known he had PKD. Maybe I shouldn't have kids either."

"I don't know," she said. Duffy's deep, empathetic gravitas radiated across the room and into me. "I think you might change your mind."

The words felt as if written over us in starlight. Her confidence that I was worth having children was so palpable, it was like an embrace. I hadn't felt anything like that since Lois died. I looked up at her; the cold glass felt heavy and unnatural in my hand, like another body was holding it while I floated a few inches above, like the opening of the vortex I'd felt on hotel beds in South Dakota on the drive east with Heather.

"After all, you wouldn't want to never have been born, would you? Or your dad? You wouldn't want him not to be here? Would you? I think you might feel differently one day."

I thought of my father and me standing beside Umpachene Falls in the Berkshire Hills. There's a photograph of me from a winter day there when I was six, in a coat with seventies white fake leather and embroidery and a white fur trim around the hood. The falls are at my back. I am beaming at my father. There is also a picture of him that I took. He stands on a shelf of rock, the waterfall behind him, dressed in jeans and a plaid work shirt, his hair still wavy and dark, before PKD took away the curl. He is blurry, at an angle to the frame, but he waves at me behind the camera.

The Berkshire Hills were once as tall as the Rockies or perhaps even the Himalayas, ground down by millennia of glacial activity until they resembled the round, friendly contours of the Shire in *The Lord of the Rings*. Umpachene Falls is tucked into a public park down a dirt road that was once a private landholding. They come down from an elevation of two hundred feet, gathering in shallow pools at regular intervals where people sunbathe or swim or erect cairns of smooth stones from the banks. The paths that line the edges of the falls are set into a pine forest that always smells fresh and wet, the paths covered in pine needles and criss-crossed by roots. At the bottom of the falls, the Umpachene River dumps into the Konkapot River, and the Konkapot broadens out next to a green lawn with willow trees lining it, sand-bottomed and full of minnows. When I was small and had strep throat in the summer, my mother took me to sit on the sandy bottom and cool my fever.

When we moved to the Berkshires, my father fell in love with fly-fishing rods and trout streams and deer runs and rifles, his father's lessons waking up in him. He made friends with men who knew how to smoke meat and fish, took me to target-shooting ranges for turkey shoots, where the prize for a bullseye was a frozen turkey that you could save for the holidays. He took me to river cleanup days, black garbage bag in hand, picking up the old beer cans, pull tabs. We were never in the same river twice.

In the course of the history of trout in the United States, trout have been moved from one region to another and back again based purely on trend and taste. European brown trout first arrived in the United States in 1883 to stock our eastern rivers for colonizers' taste. Brook trout, my favorite, orange-bellied and joyful, spotted and fierce and fast are the natives of my native rivers. The introduction of the European brown trout meant that brookies began retreating from their native streams to colder, safer areas where the browns fared worse. Brookies, by turn, were sent to the West, where they are now considered invasives, putting rainbows and cutthroats in jeopardy.

Brook trout can move from fresh water to salt water and back again. Unlike anadromous fish that spawn in fresh water but live in salt water, brookies are diadromous—they spawn in fresh water, but move into salt water only for food or refuge from inhospitable temperatures. They can be found in all the rivers that feed the Sound and the Gulf, meaning that even on Monhegan, we were not far from the fish my father loved. Measures that support brook trout of all kinds benefit other spawning fish that form the backbone of their ecosystems. Fish ladders installed in formerly dammed streams, or the removal of dams altogether, makes pathways for spawning alewives and menhaden and salmon, as well as American eels.

The brookies needed protection, return to home, support. They deserved life. And so did we, standing on the edge of Umpachene Falls, father and daughter, taking blurry photographs with a cheap camera.

Over the week I stayed with Duffy and Jack, we would learn that the CMV had become the free-floating lymphoma that invaded the soft cavities of his belly and lungs. It replicated until the system was swamped with the planktonlike cancer cells and his organs began to fail, beginning with his second transplanted kidney. He had hepatitis C from the first transplant, a mild form of polycystic liver disease, those strokes, but he was working until the lymphoma came. The fall off the cliff was so swift that I scrambled. He was the first patient his care team had ever seen with his type of free-floating lymphoma. Now it is understood as a condition called post-transplant lymphoproliferative disease (PTLP), the production of lymphomas in patients with suppressed immune systems, of whom transplant patients are a huge cohort. Years later, his case would be written up in the medical literature. But then, we didn't understand why there was cancer racing through him. And so, as the seas rose again in him, I held him as they drained the fluid from lungs or belly. We were swamped by the dark mystery of the body, blazing trails, arrested in the progress of our lives.

He only left the hospital for a few days in his last six months, each time, to be rushed back with more fluid to remove. I flew back every weekend, failing at my job, spending money I didn't have, sitting in the hospital, sleeping in the yellow room at Jack and Duffy's. When the cancer filled his abdomen and lungs again, and he had to have a huge needle inserted into his lungs to drain bags and bags of cancer-filled fluid, and he had to be helped from bed, I held him, looking into his eyes to distract him from the needles. They took him off immunosuppressant drugs to slow the cancer, and he went back to dialysis.

"Your dad's really sick?" Lawrence asked me back in Chicago, a month before my father died. We weren't playing music together anymore. Wounds were still too fresh, even after two years. He had become too connected with music for my comfort; it was difficult to do much in the city's music community without running into him. This time, we were at The Hideout, where he'd just started bartending and booking. He'd started the job when Lounge Ax, the center of Chicago's underground music scene, had closed that winter. He'd been the booking assistant at Lounge Ax since we split, and it was strange to see him behind a new bar. Everything felt like it was shifting.

"Yeah," I said. "I'm going to Boston tomorrow to see him."

"How bad is it?" He stood behind the bar and turned his deep caring on me.

"Bad. My dad is not OK. Maybe a few months left."

"Shit."

"Yeah, I've been back and forth all summer."

"I noticed you weren't playing."

"No. Gotta take a break."

"Let me know what happens. And when you're ready to play again."

Tall and broad-shouldered, he looked like Mr. Clean with his shaved head. He looked like a Viking or a lumberjack, all masculinity. But his kindness, the core wound of his mother's early death, sat just behind his eyes.

"Promise me you'll stop dating men with dead mothers," my mother had begged me after we split.

Like Ben, Lawrence was all goodness and empathy. Unlike Ben, the tragedy of his early life came with poverty, with a fierce, all-consuming need to make art. I understood that. I lived it. We didn't fail because we didn't love each other. We failed for harder reasons than that. I couldn't stop poking at the wound, even though I knew it was time to move on.

I met Lawrence in Chicago four months after I'd seen the ocean of stars in South Dakota. The first time I set foot in the city was just days after my heart had been changed by the enormity of the vortex in the Badlands. I drove into the grid from the prairies that surrounded it. I reached the lake and felt like I'd wandered into something as holy and vast as the mysteries of Sage Creek, only in city form. I left to go home and see my parents, but I couldn't shake the city. Months later, I turned back to visit it again, to see if the mystery of it was still there. I drove along Lake Shore Drive in the October sun and realized I hadn't lost the feeling. Later that night, my friend suggested coffee with Lawrence.

"I'm in love with this city," I told him only moments after we were introduced. "It's such a surprise. I'm from the hills, and I love the ocean, and Chicago doesn't have any of that, but I can't get over it."

"I mean," he said, "why should you? The music and the history and the books that have been made here? It is pretty special."

"I know, every turn I take feels like I'm in the Haymarket riot."

"Just wait. It's even better in the winter. In the winter, the lake steams and there are ice floes."

The next night we went to a bar with my friend and found a jukebox and played old country music, and Lawrence grabbed me in that unremarkable bar steps from Wrigley Field and launched me into a swing dance, the kind I'd been learning on lonely nights in Seattle when I'd gone to the queer country bar to remember that my body was also for joy. The next night, my last in the city, I told him, "I have to come back. My friend Heather wants me to pick her up and move to Portland, Oregon."

"You don't want to move to Portland."

"I don't want to move to Portland."

"Your car won't make it over the mountains."

"My car won't make it over the mountains."

"I'll write you a permission slip to move here instead." We kissed under the El tracks.

I was living in Chicago two weeks later, permission slip tucked into my wallet. In six months, we were living together. We found an apartment over Tuman's Alcohol Abuse Center, a bar where bike messengers and artists played the jukebox of indie and punk records loud and the sound of it came up through the alley into our bedroom window every time someone opened the men's room door. Junkies shot up on our stairwell, and we turned our stereo up to fall asleep to the Dirty Three and Richard Buckner, making love in Lawrence's vintage iron bed with the pleasure and transcendence of the first sex I'd had since diagnosis; touching him took me back to the whirlpool, lost in spirit and future. When we were done, he'd take records off the shelves and play me things I'd never heard.

Music was all I ever wanted. It was every desire I'd ever had, for something larger than myself, for a place to hold the whirlpool, the vortex of god that could grip me, joy in my body. I bought an ancient lap steel guitar and wrote songs in the apartment when it was quiet, and the bar was closed. I sang onstage, and my lyrics could have been about love affairs or actual glaciers. But they were about the dying body every time, and about the hope of trusting a partner to understand, while still keeping silent. They were about impossibility. And Lawrence was the way I'd linked my impossible desires: for a body I could love, for sex to take me out of it, for joy and a life where I wasn't going to die from PKD, for music, for marriage and a future.

We stayed together for a while after my proposal in that rock club, but we never got to the conversation where he told me why he couldn't marry me. I could only guess, and every guess said PKD foreclosed a future with him, my illness too big to take on. His mother died when he was thirteen. She took him for a picnic on Bear Creek near their home in Denver, Colorado, as an early fourteenth birthday gift. His best friend had a new raft they wanted to try out. They didn't know that snowmelt made the creek high and fast. Lawrence and his friend took the raft to the river alone.

They were dumped out as it flipped over in the rushing water. His mother watched, horrified, from shore. In the chaos as Lawrence and his friend found their way back to safety, neither could see her. Where she had been standing, and all along the water's edge, the riverbank had collapsed into the creek. Lawrence could not find her; then he saw the helicopter, coming to take a drowning victim to the hospital. He didn't know it was her, except in that tiny space of intuition where he wondered. She stayed in the hospital in a coma, a ventilator breathing for her, for days. He was not taken to visit her. The ventilator was turned off shortly before his birthday. He was taken to the funeral, but not to the burial near her family home in Kansas. Perhaps it shouldn't have surprised me that he wouldn't want marriage and children. Some combination of the damage from his own family, my new diagnosis, my too-frequent rage and wounded flailing in our relationship, and his single-minded need to play music cut off the possibility I'd finally begun to think I might desire.

Now, all these years later, I left Lawrence at The Hideout, left Chicago yet again, flew back to Boston. I sat in the hospital. I held my father's hand.

"I can't do this anymore, honey," he said. "I just can't. The cancer just keeps coming. Is it suicide if I let this go? If I stop dialysis?"

"It isn't suicide. It is deciding what treatment you want. You already said you don't want chemo since the cancer's moving so fast."

"But if I stop the dialysis, how long?"

"The doctors said a week."

"What will it be like?"

"Like falling asleep. Like you get more and more tired."

"But I'll be myself?"

"Yeah."

"And I can go home?"

My heart hurt. My father had moved to Boston after my mother left him. He rented a room with a horrible man, in a third-floor apartment in Arlington. While he'd been in the hospital, his roommate had stopped letting me get the mail, stopped returning my calls to get clothes, then told me to move my father's things out of the apartment. Jack and Tommy and Duffy had helped me. Everything he owned was in storage. He couldn't die in his own bed.

"No, not home. But there's a hospice room in Cambridge."

"Is there a window?"

"I went there yesterday. There's a window."

"That, then. Let's do that."

"OK."

"I feel like I failed."

"Failed what?"

"Everything. Like I'm a failure." He had no property. He had no will. He had no money. He had always tried to make something beautiful that stayed, to make lasting art or craft or to stop time and live longer, even a little longer. He'd pushed every day to make up the lost time of the illness. He knew he wouldn't be able to. And no amount of telling him he was leaving the world to me would make up for leaving the world behind.

"Put me in a trout stream," he said. "Then have a party."

When we got him into the hospice in Cambridge, I brought a photograph of him fishing for trout in his favorite stream. I brought a rosemary plant to put in the window.

"That smell reminds me of the honeymoon in Greece with your mother," he said.

I brought him ice cream. He ate with one hand and held mine with the other, his hand shaking the spoon up to his mouth.

"I can help you, Dad."

"No, I want to do this for myself." He took a bite and smiled. "Can you bring me clam chowder tomorrow?"

I nodded. Then waited a moment. "Dad," I said, "remember how you told me you were a failure? You said you were a failure, that your life was a failure?"

"I don't know, honey."

"You did say it."

"Maybe, sounds right. I never made anything that lasted."

"That's not true." He waved my words away. I gripped his hand harder. "You've had an amazing life. Your life has been incredible."

He looked at me, and in his eyes was an ocean of things he'd never had the chance to do. He'd never left the country again after his honeymoon with my mother. When I went to the town in Ireland where his grandparents were born, I'd called him in Massachusetts. "I'm in Kiltimagh!"

"Good girl!" he whooped. "Now, go get a Guinness for me!"

He'd never been fishing for trout in Colorado or Montana or New Zealand or the Rockies. He'd never gotten my mother back. He'd never seen me play. He would never meet his grandchild or go to my wedding or read my book. He'd never know I'd spend a lifetime trying to be as effortlessly cool as he was or as able to find the right people to know in any place, whether it was waiters or cooks or writers or artists. He'd never see me face my disease as it came for me. He knew he was losing all of that, while he held my hand.

When we lost him, it was raining in Boston. We were all getting ready to go see him at the hospice, making breakfast when the phone rang. Then I was standing in the kitchen in Jack and Duffy's house. On the phone with the nurse, I looked out the window. Tommy and Jack were standing with me; Duffy had a dish towel in her hand, the shady yard visible, the cold rain on the window. As we drove to see his body, I heard on the radio that in the Berkshires, his hills, the rain had been snow.

We scattered his ashes at Umpachene Falls, Tommy and Jack and Duffy and my mother and me, walking into the forest in October, a year after he died. We poured a bottle of scotch and a bag of his dust into the water and watched it move down the river to the Konkapot, where it would join the Housatonic and then the Sound and then, at last, the Atlantic. All I wanted was another second with him, and it didn't matter that the second I'd get back was one where he and I both had PKD. Duffy was right, I felt differently, that I was worthy of being alive, and my father was too, and any child I had would be as well.

He was always the father with the long, dark ponytail, the one in the pictures where he holds a film camera up to his eye, where he holds a hammer over his head in a gesture of glee, where he flips off the photographer, where he holds me in a backpack carrier, and we walk through the snowy forest for a Christmas tree to cut down. He is always in the hospital bed, tennis is on, Tommy sits at his elbow with the *Times* in his lap. He is always leaving the library with a stack of plastic-cased hardcovers from the waitlist. He is always reaching down to show me how to take a fish off a hook, how to aim a shotgun, how to weave a basket handle, how to plane a plank of maple. He is always sending me newspaper articles and listening to what I love.

The last night of his life, in the bed in the hospice in Cambridge, a slice of blueberry pie on the bedside that I'd brought him, he was holding my hand as the visiting hours ran out. Our family was upstairs, they'd made dinner and sat with him, cleared the dishes in the common dining room of the hospice. They left us alone like that, in that room, the streetlight outside the window, the rosemary on the windowsill. He was in and out of sleep and the morphine was strong. They dosed him again after his meal and he drifted.

"I wish we could go sailing," he whispered.

"Me too, Dad."

"I love you."

I have not lost him, ever. He is always paying attention. And even though he is not here to talk to, and even for all the rage he had, for all the failure he worried about, he would have been with me in any hospital room, and he would have known exactly what I was feeling, like no one else in the world. I understood how to be the child in a hospital room with my father, sitting in the uncomfortable chair, holding his hand. I did not understand at all how to be the parent in the bed, the child at my side.

I did not understand until that day in Maine, alone with Dex—trying to be ill and somehow to be in charge, trying to be safe and somehow be in danger—how hard it must have been for him. I'd resented how he hid himself and his illness from me. I'd resented how he would disappear into fishing, into making things, into reading, always moving so fast to learn more, make more, understand the Konkapot, the Housatonic, the Sound, the fish. But in the bed, I saw how it happened to you. I wanted life more than anything. I wanted to be a parent more than anything. I wanted to make art more than anything. And some days, most days, when I could tell myself my illness wasn't relevant, I would grab for that, for normalcy, for passion, for joy. He had done that. And when his illness came to claim it all back, it looked like failure, like the loss of a habitat or a species or a future or a relationship. But it wasn't a failure. Life itself, his and mine, were worth saving, worth having, no matter what.

Common Periwinkle

(*Littorina littorea*)

In the Connecticut River Valley, my mother was packing. She packed a bag full of clothes and one of knitting and walked from her subsidized housing apartment in an old Catholic school through the quiet downtown of Northampton and got on a Peter Pan bus bound for Boston. It was a two-hour drive to Boston from where she lived, but the bus took three. She called me from the road. "I got on the bus!" she said on the phone while I was still in the hospital.

In the early morning, the nurse had come in, not waking Dex, "You've been able to stand up without passing out. And no seizure. So, maybe you really don't have a seizure disorder. With follow-up, they'll figure out what happened."

"OK."

"And you have a plan?"

"I think so."

I'd rented the very cheapest house on the island for the very shortest amount of time. Dex and I looked at pictures of the house online, a tiny place close to the meadow wetlands at the heart of the village. It looked run-down but cozy.

"Holdfast House," he said, "like your record."

"That's right." That year I'd put out a record, my third. Dex had come with me to the studio, to band practice, to shows, *Slipping the Holdfast*, named for what sailors tell one another in a storm, *hold fast*, named for the kelp forests of California, the holdfasts at the base of their long bodies. At the end of the season, the kelp let go, releasing from their anchors, and take to the open sea.

"OK." The nurse whisked into the room. "Do you need help packing?"

"We didn't come with much," I told her.

"Well, you can't forget that stuffed octopus, right?"

Dex woke up at the word. "No," he mumbled, groggy. "That has to come. That's my friend."

After breakfast the staff wheeled me out to the cab. Dex rode in my lap. I was dizzy enough to be grateful for the wheelchair, my blood pressure was still dropping every time I stood.

I rolled down the cab window, and the cool air brought me momentarily awake, then a wave of vertigo engulfed me and I closed my eyes, rested my head against the back of the seat, and fell asleep on the way to meet my mother.

Monhegan was still waiting, our refuge, a scale we could handle, where a child with a bucket is the richest person on Earth. Its wonders are close to all the wonders of the Maine coast. Even the license plates declare the state "Vacationland," ignoring the complex history of place, the class struggles, the Indigenous lives, the factory towns, the logging scars, the warming water, the pollution history as complex as that of the Sound. Maine's most famous science writer, Rachel Carson, from her home on Southport Island, wrote books about the complexity of the regions where land meets sea long before she became a prophet of collapse.

Her early works were lyrical meditations on the tide pools of Maine and the oceans everywhere. She had a gift for bringing people into alien landscapes and serving as their guide. In *The*

Edge of the Sea and *The Sea Around Us*, she took readers by the hand and showed them beauty, ecosystems that were healthy—mollusks and algae and fish and seaweed. She revealed complexity as harmony. Many of her early readers balked at the stern warnings that laced *Silent Spring*. But, for Carson, there wasn't much difference between enumerating beauty and enumerating threat. When you saw the wonder in something, her books implied, you would naturally want to preserve it. Her last book, *The Sense of Wonder*, remained unpublished at the time of her death.

In *The Sense of Wonder*, Carson wrote, "If I had influence with the good fairy who is supposed to preside over the christening of all children, I should ask that her gift to each child in the world be a sense of wonder so indestructible that it would last throughout life, as an unfailing antidote against the boredom and disenchantment of later years, the sterile preoccupation with things that are artificial, the alienation from the sources of our strength. If a child is to keep alive his inborn sense of wonder without such a gift from the fairies, he needs the companionship of at least one adult who can share it, rediscovering with him the joy, excitement, and mystery of the world we live in."

This was the project of my days at the tide pools with Dex. It was at Monhegan's tide pools that we learned to sing to periwinkles. If you pull them from the water, they will tuck their feet inside their shells. But as they warm in the sunshine, they begin to reach out and explore. If, at that moment, you sing, *Periwinkle, Periwinkle, open your eye/Periwinkle, Periwinkle, don't be shy*, they will come out even further.

The tourist who taught us the song was no more a scientist than I am, but she made a convincing case. "The vibrations from your voice," she said, "or maybe the air from your mouth, it makes them curious."

If you find a brave snail, their foot will extend and the creature will reach and reach, touching your rough thumb with its foot, then extending farther until their eyestalks emerge into the sunlight, and, looking like they might be blinking, they seem to regard you and your strange music with skepticism. It is a Beatrix Potter trick of imagination and tune. It is a miracle.

For people who casually spend time on the coast of the Atlantic, periwinkle snails are as familiar as the sound of the waves. There are three species on the Atlantic coast. Two are native, the smooth and the rough periwinkles. The smooth prefers the sea, the rough the land. But the most plentiful on the Eastern Seaboard is the common periwinkle, *Littorina littorea*, straddling the worlds of land and sea. About an inch-and-a-half high at maximum, gray-brown-black, *L. littorea* turn chalky when kept out of water. Conical and whorled backs have a dumpy base, the classic point, the lip of their opening a shiny white, their soft parts dotted with a punctuation of dark bands. Periwinkles are used as food in Europe. Periwinkles may have the same name as the little blue flower, but likely got the name because they were available for a penny a piece, or a penny a *winkle*, the common word for snail.

The common periwinkles are invaders. They may have been brought to the colonies as an easy source of protein. They also may have arrived in the ballast of ships, on the ballast rocks that weighed down empty holds until cargo took the place of discarded stones.

Where they live in the intertidal zone, common periwinkles can be submerged or semi-terrestrial. They lay their eggs in the sea, though, and hover in those spaces at play in the tides, bathed by the rhythm of the ocean. They eat algae—grazing over rocks, consuming the coating of plant matter that covers them, changing the nature of the ecosystem, stripping rocks bare, revealing their gray surfaces. They have limited natural predators to slow their destruction of ecosystemic balance. Their grazing mechanism, the tonguelike radula, works its 3,200 chitinous teeth across the rocks relentlessly, an army of snails eroding the coastline alongside the wind and water.

Periwinkles feed in marshlands, breaking the stalks of cordgrass until those wetland anchors die. Smooth cordgrass is the frontline defense of the Gulf's salt marshes—an inland ecosystem of marsh and meadow that acts as nurseries for shellfish and fish and shorebirds. Without the strong anchors of cordgrass, wetlands on the edge of the ocean cannot face the strong currents of the Gulf of Maine. When cordgrass dies, salt water floods the ecosystem, upends its delicate balance of salinity, and erosion takes that wetland back for the sea. They eat barnacle larvae, disrupt diatoms—that planktonic anchor of the marine food web. Diatoms are in a class of life responsible for more than a quarter of the photosynthesis on planet Earth, threatened by ocean acidification that weakens the tiny transparent shells that enclose them like greenhouses. In 1992, James T. Carlton, in "Introduced Marine and Estuarine Mollusks of North America: An End-of-the-20th-Century Perspective," concluded that "no introduced marine mollusk in North America has had a greater ecological impact" than *L. littorea*, which "altered the diversity, abundance, and distribution, of many animal and plant species on rocky as well as soft bottom shores." By now, they may have been eclipsed by the zebra mussel, but that took some doing.

We sing to them still, aware that they are dangerous, in awe of their lives.

Rabbi Abraham Joshua Heschel wrote in his book *Man Is Not Alone: A Philosophy of Religion*: "Wonder is a state of mind in which ... nothing is taken for granted. ... Each thing is a surprise, being is unbelievable. We are amazed at seeing anything at all; amazed not only at particular values and things but at the unexpectedness of being as such, at the fact that there is any being at all."

I woke up at the driveway of the Gosnold Arms Inn, the hotel where we'd left the car on the morning we'd taken the ferry over to Monhegan on the day of the accident. Dex and I walked up the long driveway, up the hill to the cottage room that had been the last one available that night. In the morning, we'd take the car to another ferry, the last one with tickets left, and then get back to the island. I could feel my mother's travel. My phone had sketchy reception, but on the tarmac, I saw she'd left a voicemail message: "Made it to Maine! I'll get off in Damariscotta. I think I have to get a cab. See you soon!"

I felt the common wave of hope in my chest, hope she'd arrive and tuck me in to bed, take Dex out for dinner and bring me soup. Maybe this time. Maybe being in the hospital was enough. We walked to the cottage and unlocked the door, then lay on the beds. My head was spinning; Dex climbed off his bed and curled next to me. If my fall wasn't a seizure from a seizure disorder, and it wasn't an aneurysm, and it wasn't a stroke, maybe it wasn't PKD causing the accident, or maybe it was some new kind of disordered system inside the illness.

I thought again about my Uncle Brian, my father's only sibling without the disease.

In 1990, Brian was told by a doctor that there was no chance he had inherited PKD. He was forty-three years old, uninfected by HIV still, and the age at which his family was already very ill or dying. He was released from the curse, from the single factor that had determined all his decisions. Brian's longtime friend Don told me, "Brian . . . used PKD as the reason not to do certain things." He'd avoided commitment, career, savings, retirement planning. His story mirrors those of HIV survivors in the later years of the plague. As new drug regimens extended life expectancy, HIV+ patients came back from the edge of death. As Don's partner, Gary, put it: "You get it, you die . . . people would leave their jobs . . . do all their travel figuring, I'm gonna be dead." Then, "all of a sudden . . . by the way, here are these new pills. You're gonna survive, you're gonna live. But people had no more money. They had no job. They were letting go of their apartment, and it's like, guess what: Now you have to reconstruct every little thing."

Standing in Don and Gary's kitchen, just after finding out he didn't have PKD, Brian told Gary he felt the same. It had been eight years since the first man in America was diagnosed with HIV. In a speech, years into the crisis, Larry Kramer, playwright and HIV/AIDS activist, said: "Don't you ask yourself quite often the big question: 'Why am I still alive? At some point I did something the others did. How have I escaped?'"

What must it have felt like to stand on the edge of death for so long, surrounded by the dying of two families, the water around you, the last breath in your lungs, and then to blink and find yourself facing a placid shoreline, the waves far away?

There is a great deal of writing on survivor syndrome among the gay men who outlived their chosen families, on survivor guilt in veterans, Holocaust survivors, refugees. But somewhere in the years after he knew PKD wouldn't kill him, he did acquire HIV, and left the world of survivors. None of his friends know how or why. But maybe the fairy tale had something to do with it. It is tempting to reduce Brian's story to any one of those categories, and perhaps it borrows something from all of them.

But to Gary, the idea that he intentionally acquired HIV is impossible. "I'll tell you that, after he realized he didn't have kidney issues, I asked about HIV. And he said: 'I don't think I'll get it. I won't have HIV because I've been very careful. I don't do any of those things.'"

For at least 150 years, PKD had caught everyone, drowned everyone, killed everyone—as if we lived in a fairy tale castle that filled up every night with the ocean, while we tried to hold our breaths until the morning. Only one of us survived the curse, but the curse in his chosen queer family also drowned a generation. He kept a record of that loss—a magic object, a book. I saw it in the apartment where he lived, on a small street in the center of gay Philadelphia, in a row of narrow brownstones, in his world of beautiful objects. The dresser in his bedroom was painted to look as if made of malachite; a gold Dunhill lighter was in his pocket, a silver cigarette case on his coffee table painted faux marble, an antique Chinese jar on the table where he wrote, full of pens, decorated all over with drawings of twining dragons and fish. Next to the jar, there was an unremarkable black address book.

I picked it up, held it, turned it around.

"That's my most precious thing," he said, taking it out of my hands.

"Really?" I asked.

"That's the book of the dead."

He opened the book and flipped through the pages. They were crowded with the names of people he'd known in the whirlwind of gay 1970s and 1980s New York City and Philadelphia. When he paged through the Cs, I saw the name and address of my father, Charlie Caffall, Brian's brother, who was still alive then. Next to it were the names and former addresses of my Aunt Lois and Uncle Patrick, my grandmother Kitty—all gone. Next to each a little tick mark. As he turned the pages, I noticed tick marks next to almost every name.

"What are those?" I asked.

"I put a check next to the names of the people we lost to AIDS, to kidney disease." He stared at the book for a long moment. "I can't bear to erase the names of the dead."

Brian was behind the scenes during the crisis and its activism, working as a writer. If there was a need for advertising copy for the AIDS Coalition to Unleash Power (ACT UP) to use on condom PSAs in Philadelphia, Brian was handed the raw copy and made it better. He did the same for sex education pamphlets handed out at bathhouses, and for posters for fundraising balls. As the crisis went on, he transitioned to writing about gay life for the *Philadelphia Gay News*, where he had a regular column called "Pith and Vinegar," and for other newspapers, nominated for awards and keeping it light, focusing on the joy of the community as often as not, like he'd been taught as a Caffall, not dwelling too much on the pain. One of Brian's oldest friends, Bill, told me that all of it mattered; his writing, "his contributions were extremely important because he was able, with words, to reach people. You probably have that same gene."

When I was diagnosed with PKD, I went to visit Brian. We sat on the pink couches in his apartment, tea on the coffee table painted to look like marble. He played me a cabaret musician on CD. "I saw her sing in person last week," he said, "not a dry eye in the house." The song ended and a quieter one started.

"I wanted to come here, to tell you in person," I said. "I can't imagine telling you on the phone."

"What, honey?" He turned the volume down on the stereo, picked up his teacup.

"I was diagnosed with PKD last month. I had an ultrasound. I told Dad. I asked him not to tell you. I didn't want you to hear it before I could see you."

All the color drained from his face. He put the tea down and put his head in his hands. We sat like that for a long time.

"Are you feeling OK?" I asked. "Dad started drinking again. I told him last week, and he told me, 'You aren't the disease,' and then just didn't talk to me for a week. I can't get him to say how he feels."

Brian still didn't look at me. He kept his head in his hands. "You'll be OK."

"That's what Dad keeps saying."

"You'll be OK."

Outside the sun started coming down. The CD he'd been playing finished. It was getting dark in the apartment, and I could see the ways in which he'd let it be dark. He didn't have many lamps; curtains sculpted the light from the street. I could see the bottle of Glenlivet, the glasses close by, the books everywhere, the writing table, the pens in the jar, the address book lying next to it. My name was in there too.

We sat there like that for a long while. Then he picked his head up, walked to the table nearby, poured himself a drink, and said, "Let's go to dinner."

"It was huge," Gary said. "Yeah. It was huge. When you were diagnosed. It was just a sword of Damocles. It was devastating for him."

I was trained to live to tell the tale, Melville's Ishmael speaking the names of the dead. Brian could only see himself through the curse. It is important to say that he did not bring on his HIV infection. I have been sick with a chronic illness for a long time. I feel only rage at any implication that the sick bring illness on themselves or think their way into cure. I am not looking to be cured, I can't be, I am looking to be healed. The former is about science, the latter is about belief. The former acknowledges a disease system that operates through biology, the latter allows for complexity in the quality of my suffering. The world is too random, and the heart is too complex to assume that we have control over our bodies, and illness is not a judgment, nor is any disability. It is chance and genes and pollution, it is stress and poverty and capitalism and luck. It is not you. It is not me. It was not Brian.

That he contracted HIV was not his death sentence, any more than my PKD is mine. People survive both curses every day. The story he was told about his future as a Caffall and the country in which he lived might have been. His lack of planning meant that he didn't have any savings to afford health care. The terrible state of health insurance in the United States meant he had few options as a person living in poverty. When he showed alarming symptoms, Don and Gary and other friends insisted he visit their friend Dr. Nick Ifft, coordinator of the Philadelphia AIDS Task Force, who saw Brian for free and connected him to other free medical services. He diagnosed Brian as HIV positive. By then, his disease was too far along for the cocktail of drugs that could have restored him. He had HIV-associated dementia, lung cancer spreading to his brain. Slowly the castle filled with ocean again, every night, while he sat alone with Glenlivet and cigarettes, the curtains drawn on the darkening street, and maybe he practiced imagining what it might be like when he failed to hold his breath. The ocean rose over his head.

Bill told me, "You know, when he found out that he had HIV, he made a chilling statement to me, 'I now really feel like I'm a member of my own family.'"

Brian died in 2003. He was fifty-six. He was cared for by the infrastructure created and accessed by HIV/AIDS activists—Gay Men's Health Crisis in New York, Meals on Wheels, HIV/AIDS buddies—a future they made for themselves of mutual aid and survival, a future they imagined and enacted in the middle of a curse, a plague.

On that visit when I first saw the address book, Brian read to me from his favorite book, James Agee's posthumously published masterpiece *A Death in the Family*. In that book, a child loses his father at six, the same age Brian was when his father died of PKD. In the prologue, "Knoxville, Summer 1915," the child voices a memory of safety and peace. "'By some chance, here they are, all on this earth, lying on quilts, on the grass, in a summer evening, among the sounds of night. May God bless my people, my uncle, my aunt, my mother, my good father, oh, remember them kindly in their time of trouble; and in the hour of their taking away,'" Brian read to me.

Was it any wonder he came close to the family and then ran away? Was it any wonder that he brought gifts, and kissed us all on the forehead and looked at his siblings, failing and exhausted, and ran back to the people who he'd chosen, people who were also beginning to die?

I understand the urge to look only at the end we face, and not at the amazement. I see the periwinkle as the harbinger of coastal destruction; I hate it for the ruin of the marshes, fear the slow work of its radula on the rocks and algae. But, as Carson said in *The Sense of Wonder*, "I sincerely believe that for the child, and for the parent seeking to guide him, it is not half so important to know as to feel. If facts are the seeds that later produce knowledge and wisdom, then the emotions and impressions of the senses are the fertile soil in which the seeds must grow."

It is impossible to sit by the water watching a child collect snails as if there is not a collapse unfolding. It is impossible to rely on wonder alone. It is like living in a body that is shifting under your own skin, out of balance and beyond the control you thought you had. My parenthood, that night on the cutter, those days at the tide pool, was a project of holding two things in mind at all times, and training Dex to do the same. It felt impossible sometimes, as if only rejecting my disease would keep him safe, pretending it wasn't waiting for me and maybe for him. It felt impossible to avoid some days, days when his father starved us of time and money out of anger, punishing my vulnerability and by extension Dex's. But holding both was the only way forward I could see. Wonder is only a step away from denial. The spiral of mourning requires seeing both. We should take strangers by the hand, stand with them in awe of what we have, and plan what we'll do to make sure that it exists at all.

I lay in the quiet of the little cottage room at the Gosnold Arms Inn, Dex against my chest, my heart thrumming. *What would I do if I could heal*, I wondered. What would it look like to let go of the curse at last, to turn away from the panic of the divorce, to let people help me, and to picture a life where I survived the flooding, the terror? I let myself hope for care from my mother. I pictured her travel like the old maps in classic movies, the bus with an arrow moving over the watersheds that fed the Sound, and then fed the Atlantic just off the borders of the big city we never went to. And there was my mother, in pain, not used to traveling anymore, sitting in the stale air of the Peter Pan, asking strangers how to transfer at the station to the bus to Maine.

"When is Grandma gonna get here?" Dex asked, draped over me in the hotel room in Maine.

"I don't know. When her bus gets in, I guess."

In the room at the Gosnold Arms Inn, where I could smell the sea, where I knew my mother would stand on the deck and say, "Smell that!" when she arrived, I sang a lullaby to Dex to pass the time, Fats Waller's *Hold Tight, hold tight, hold tight, hold tight, hold tight/Foo-da-racki-saki/ Want some seafood, Mama.*

Comb Jellyfish

(*Ctenophores*)

Dex waited on the porch at the Gosnold Arms Inn, looking at every car for my mother—waiting for one of his favorite people, an adult to help us after the hospital and the night of the apple tree. It took an hour, and when he saw her getting out of the cab, putting on her field hat, he ran to meet her.

"Well!" she said, walking up the hill, Dex carrying her bag. "This is so pretty! I can smell the sea!"

Dex held her free hand and beamed at me. A warm wash came over me. My mother, arriving with rescue. I was five and hiding behind her skirts as she blocked me from the mean teacher. I was twenty-eight and brokenhearted after Lawrence moved out, and she was driving me to the Cape for a weekend together reading on the sand. I was thirty-seven and fresh from the derecho. I was forty-three and newly out of the hospital. My mother was alive; she was sick, but not with PKD. My father was gone, and in the last years, I was his rescue. He was so close to me, so like me, that, especially since my diagnosis, I made him the hero of every story. But there was my mother. She was here. I felt myself soften, ease into my love for her, walking up the road, holding Dex's hand, field hat on her head.

"You made it! Thank you!" I called from the porch. "We're in here."

"Mom! Can we take Grandma to the salt pond?" My heart fell, I was tired.

"Maybe, kiddo, if she can drive."

"Well, let me get settled and we'll see!" She leaned down and kissed Dex on the head.

"We can go to the salt pond, right, Mom?"

"Can you drive there?" I asked my mother.

"How far is it? What is it?"

"It's a huge salt pond, right on the coast, right off the road. We go every year before we get on the boat. Supposedly Rachel Carson did observations there for the book *The Edge of the Sea*. It is really pretty. Could you take Dex so I can rest? Only a quarter of a mile."

"I need some rest too, so if I can lie down first, OK. But I can't drive alone. I haven't driven stick in a decade. You'll have to be there."

"You'll have to get all the way to Massachusetts in a few days."

"Sure, but you'll be there the whole time."

"Yeah, I guess so."

"I have to practice sometime."

"OK."

She was so like him, twinned the way she had been to Mac— science mind to science mind, unique mind recognizing itself.

After his father moved out, Dex slept in my bed every night. I'd fall asleep with him, singing about fish, then wake an hour later, a list of unpaid bills and unwritten books and unsung songs pounding in my brain. I remembered what his teachers said at school.

"He's stopped relational play lately, just plays alone, not even parallel."

"Is that a problem?"

"It's a regression."

"The divorce?"

"We wondered that."

With me he was the same as ever, the same as I'd been—talkative, imaginative, kind, with a tunnel focus on animal life. He loved what I loved. He was just like me, just like my mother.

In the car, driving anywhere, he'd tell stories from his car seat. "There are giant spiders following the car, Mom, and they have called out to the trees and the trees are all singing in the wind language." He'd make movies on my phone about dinosaur poop and turtle kings and cradle tiny snakes in his bare hands and hold them up to me as prizes, talk to strangers about blind cave crickets and glowworms and the dangers faced by bats in regions affected by white-nose syndrome. He watched hours of *Planet Earth* and recited stories of cave swiftlets and *Vampyroteuthis*. We played a game whenever we had to wait for anything, one he made up. "I have an animal in mind," Dex would say, then roll out clues only I would understand until I guessed the species.

Once when I'd come to pick him from preschool, his teacher told me, "He made up a job today during creative playtime. We put down a kiddie pool as a pretend skating rink and the kids were skating on it in their socks. Dex made a pretend hot chocolate stand next to the rink. I looked over at one point, and he was crying. When I asked him what was wrong, he said he'd run out of hot chocolate. But he told me, 'Don't worry, my assistant is getting me more.'"

This was the same child who fell apart at playdates because he could not forget that I had left and would stand at the door to my friend's apartment wailing until they called me and I came back from errands, from precious time alone, to take him with me wherever I went. This was the child who would never take a bottle, who would only nurse. This was the child who couldn't sleep more than two hours a night until he was seventeen months old, who dropped his naps before two, just as I had done. I remembered the battles about sleep I'd had with my mother when I refused my nap. I fought and fought. My mother resisted and refused. With Dex, I adjusted, empathized, soothed, supported. My exhaustion showed what it took from me.

When his father pushed back against the routines I created, I told him we were front-loading care and safety, and that trust, once established, would survive all the future tests of the teen years and beyond, but lost, would never come back. I knew what had broken my trust with my mother, long before the drinking and the divorce.

She insisted I hide all I needed from her because it was hard. She'd done that with my father too, failing to relinquish her own history as the sick child, the sad child, the lost child who needed more. Since no one else could have that, no one else did, and she asked for our care by refusing to care for us emotionally, even as she threw all her energy at saving the family through work and sacrifice.

But even that aspect of how she was in the face of the drowning family obscured the story. In linking so early with my father's diagnosis, in feeling I lacked the strengths of my mother and Mac, in my father dying just before I was mature enough to see them both right-sized, I'd done what so many women do—lionized my father at my mother's expense. I'd identified with his illness, not hers. I'd identified with his way of thinking and seeing the world, book-smart, creative, curious; not hers, scientific, observational. Both had asked me to care for them. I only resented my mother's request. I'd never seen how much my mind was like hers until I saw the way Dex's was. I ached because I couldn't count on her rescue, the ways in which I was forced into the role of her sister, her mother.

It was complex, what she did, crying when I came to her sad, angry at me when I was ill, but ready to sell her house to bail me out of a bad marriage. It meant there was no steady story with her. No single answer.

In my parenting I was turning the wheel in the opposite direction—money, career, these things were thrown on the bonfire of saving the dyad of understanding Dex and I constituted, building his ability to recognize what it looked like when there was trust between two people, trust he could replicate down the years in every relationship, so that he never believed for an instant that he was *less than* in a world that insisted he was.

When I told Dex that his father and I were getting divorced, he looked straight through me. The books had told me what to say to him. His father had refused to have the conversation, so I sat alone with Dex on my lap in the living room of the house we were losing. "We'll always be a family," I told him.

"Mom, count my toes," he said.

"OK?" I counted to ten.

"That's the percent chance that we'll be a family."

We went on Medicaid and food stamps while I hustled to build a freelance career. I thought about going to seminary, about graduate school, about ex-lovers who might make me feel safe. Custody conversations stalled. Dex needed support; regressions mounted until kindergarten, when we began trying to figure out the complexities of his mind. In a meeting at the school, his father yelled at the social worker, "You're trying to diagnose him so that you'll have less work to do. Or maybe *she* convinced you to do it to help her in the divorce."

The social worker, a tall Black woman who told me in confidence she'd once worked in a domestic abuse shelter, stood to her full height in that bland 1950s room in that urban school building and looked him in the eye. "Sir," she said, "this is much more work for us than you can possibly understand. We would never do this unless it was necessary. And if you can't get on board with the work, your child will suffer."

I added appointments and therapies to our routine. There were random days when he had to be taken out of school. We went into divorce mediation, where we talked through parenting a sensitive child in two homes. Working in an office slipped out of my grasp. Graduate school did too. My world narrowed down to surviving. I made more money writing at night. I left food stamps behind, but not Medicaid. The divorce let the house slide into foreclosure, and I went to the mailbox every day anticipating a letter from the sheriff telling us to move.

Underwater—that was the phrase for the years of the housing crisis in 2009, 2010, 2011—I was underwater. And it did feel like drowning. I could read books about the end of the oceans, but not about children in peril. It felt so deep that I put myself to sleep writing out the things that terrified me most until they didn't haunt my nightmares. Every day and every night I thought, *Please, let him be older before I need a transplant, let his father allow me to keep him if I get sick.*

I tried to look as if everything was fine. I worked but tried not to sound desperate when I asked around for more. I invited people over and didn't tell them we'd stopped paying the mortgage. I went to the doctor but only vaguely heard when the new nephrologist I could see on Medicaid told me things were getting worse quickly, higher blood pressure, larger kidneys. My mother sold her house, three years of negotiations and counting, mediators and *guardians ad litem* and therapists all trying to settle things, end things. *Alligatoritis.* I was nearing forty, Dex was nearing seven. I had already beaten the Caffall record as the oldest person with sick-but-still-working kidneys—I lay awake wondering how many more years I would get.

I clung to Monhegan as the only thing that mattered to us both. Dex would tell strangers about the island. I'd drive twenty hours to my mother's apartment, sleeping on the pull-out couch so that I wouldn't have to pay for a motel. I'd pay for the ferry in cash I'd earned at gigs, singing while Dex was at his father's house for the weekend. I lost my credit in the divorce and acquired a horror of debt. I thought about the years when the electricity was turned off at the house in Great Barrington. I thought about the government cheese and peanut butter we'd eaten as my father got sick. Dex was another generation on government assistance. I thought about the things I couldn't give Dex and the things I hadn't had. The picture of me in Maine that I'd seen the year the marriage finally broke, the one that looked like hope, was in a storage locker.

"Let's go see the dinosaur footprints," my mother said the morning before Dex and I left for Maine, the summer Dex was seven.

"On Route 5?" I asked. "I've never been there."

"Didn't you come with me on that geology department field trip?"

"No."

"Well, you should see them."

She was too sick to drive, so I drove us to the edge of the Connecticut River, still stiff from the long day in the car.

"There! Pull in right there." Her whole body tensed, worried I'd get it wrong, worried she'd get something wrong, anxiety that would mount as I waited for an outburst, and as my avoidance increased her frustration. We could not keep from the loop of it.

"Dinosaur footprints?" Dex asked.

"Yup. Turn! Turn!" Her anxiety like a river, still trying to get it right.

I pulled into a dirt lot, parked the car, and unhooked Dex from his car seat. She stood by the car door, put on her field hat, the same straw hat she'd worn on the field trips I'd gone on when she was in her undergraduate program. The same field hat in the pictures of the month she'd spent doing fieldwork for her master's degree in the pale sun of New Mexico, where she stayed with colleagues and may have been happier than she ever was in her life. The same field hat she wore in a book I'd made at New York University, a book on women scientists, with interviews and photographs. In the book she talked about how hard it had been to leave me at home in Great Barrington while she went to school two hours away. "I wasn't able to braid my daughter's hair," she had said to her daughter, me, interviewing her. She told me I was brilliant. She read every single thing I ever wrote. She helped me get back to school when my father got well. "Go study abroad," she said, echoing the conversation she and Mac had on the couch the night of the fifty-fifty chance, way back when my diagnosis existed as merely chance. "I'll take care of things here." She provided health insurance for my father after they divorced. She wore the field hat when she took me to the Cape for my broken heart.

"Down there, Dex," she said, and took his hand and the two of them scrambled away from me down the trail and the stairs onto the shale that flanked the river course, a mimic of the two of us on Penfield Reef. She'd given him her rock hammer to hold, because he loved it. She'd told him these fossils were just for looking. From the top of the trail, I could see him looking up at her, bouncing on the balls of his feet. She limped next to him, her pain making her stoop, her face bloated. She'd been newly diagnosed with fibromyalgia, diabetes. She described feeling pain and heat along her body on all the pressure points, as if stress and loss lit up every nerve pathway.

"Come down!" she yelled, "We found a fern!"

The ancient Connecticut River was a wetland with braided watercourses. Packs of dinosaurs left footprints until the mud set into stone, a record of pathways. The modern Connecticut had defined banks, and in this one section, shelves of shale touching the water, full of those fossils. The gray shale of the dinosaur footprints stretched to the river, in some places washed by its wavelets, with fossil imprints of mud from the Jurassic and Triassic, fossil imprints of the river's wavelets as if carved into the flat, gray rock, a record of water.

The dinosaur footprints were first found in the region in 1802 by Pliny Moody. He called the huge birdlike impressions the tracks of "Noah's Raven," thinking the footprints were of enormous ancient birds. This idea stuck with most early scientists, including the most famous collector of these imprints, Edward Hitchcock, ichnologist and future president of Amherst College.

Ichnology is a branch of fossil study referred to as trace fossils, images preserved in stone not of biological material, but of the evidence of biological activity—footprints, not feet. Most of the best imprints are in the basement of the Amherst College natural history collection in the Beneski Museum of Natural History, in a modern, elegant redesign of the nineteenth-century basement—low light and glass walls and rows and rows of hanging gray shale with the three-toed imprints of walking dinosaurs. It is the largest collection of these trace fossils in the world.

In Chicago, Dex and I nearly lived at the Field Museum. There, in a carpeted space of dark gray and quiet, is a huge octagonal display counter, inside a hundred sizes and shapes of trilobites, all arranged in their fossil forms, balled like pill bugs and splayed like horseshoe crabs, compact or enormous, detailed or indistinct, from all over the planet. Behind the display, there was a room-circling screen on which was projected a digital reproduction of the seas of the Cambrian period. We would sit on the floor, Dex cozy inside the boundaries of my crisscrossed legs, and watch the whole film. The room was dark enough to allow immersion. With the water's surface at the top of the screen, you were under the sea and there was nothing more important than the water. Water plants waved in the dappled light, and creatures floated or swam through the scene, pulsing with alien construction. In displays below the screen, placards detailed the species in the film—*wiwaxia, anomalocaris, opabinia, hallucigenia, trilobite.* The loop of the film ran twenty minutes, and we'd wait until the whole parade had gone by, Dex calling out the animals as they passed, memorizing their scale and colors, their movements, the moment when one would prey on another, until he had it perfectly preserved in his mind, a song he could sing when the radio was off.

Trilobites were marine arthropods, distant relatives of the horseshoe crabs that came ashore on the Sound every summer, covering the August sands of Dogshit Beach with their prehistoric bodies. I'd pull horseshoe crabs up by their lancelike tails and strand them on their backs to examine. Flipped over, their six pairs of legs like a Geiger monster flailed in the sun, waiting for the power of a wave or a small child to turn them over again. I'd watch them mating, watch boys attempt to get them to raise their weapon-tails by throwing stones onto their backs, where a nerve would induce them to strike the empty air.

At the dinosaur footprint site on the Connecticut, little is left, the most impressive fossils are all in the museum. What remains is mostly the remnants of plant life, ferns and cycads and pines, the tiny imprints of walking insects, the action of lost waves—the Connecticut, revealing to us the complexity of its history, the history of 200 million years ago, and the one written in my mother's limp. What was clouding my mother's body was so complicated that most days I didn't understand it at all. She was bedridden, but she had been from hangovers when I was a kid. She was exhausted, but she had been the whole of my teens. She was depressed, but that was with her for life. She'd lost so much to caring for me, to caring for my father, to caring for the land. It was hard to know which thing had broken her body. She and I had walked the old landfill site west of Great Barrington when I was a kid, digging wells to test the water for chemical leachate as it entered the nearby stream that fed the Housatonic. She'd gone on to do that for years. I imagined the things in the soils she'd tended and the rivers she'd walked, and then I heard her moaning in the night. I found it hard to separate my anger from the evidence of her illness. She wasn't angry at me. But I was angry at her.

The anger was old. The anger fought with the admiration for all she'd done. While Mac was dying, she'd boarded a bus on Sunday night to ride from Great Barrington to Amherst, spent the week in classes to get her undergraduate degree, living away from us in a dorm, then boarded another bus home to spend the weekend caring for Mac, caring for my father, caring for me. She wrote two theses, an undergraduate one on groundwater leachate from that old landfill, and a graduate one on the geophysics of a single gorge in Taos, New Mexico. She did this all with severe dyslexia. She did this with a lapsed, unfinished, decade-old education in theater from Bennington College. She went to Hawaii with vulcanologists when she had agoraphobia. She climbed mountains when she had acrophobia. She managed to see my father through his failed transplant while she was starting her first job out of graduate school. She managed to help him through dialysis while working full time as a hydrogeologist, first for a consulting firm and then for the state. She'd done all of this as a woman who was told her whole life that her only value lay in her looks, that she was too stupid to do anything but be an actor. She did this while battling suicidal ideations starting in her teens. It was no wonder that all that was left after working that hard was a blasted internal ecosystem of illness and exhaustion.

When we left for Monhegan the day after the dinosaur footprints, I took some of her geology books. Back in Chicago, I read about water pollution and soil remediation late into the night. Dex walked in on me watching a documentary about nitrogen runoff. The film detailed the impact of farming practices on the Gulf of Mexico, describing the dead zones of biodiversity left when the Mississippi River drains most of a continent's worth of fertilizer into the sea. It described how only comb jellies proliferated in the dead zones, most other fish life having been devastated by the chemical changes in the water. He crawled into my lap, mesmerized by the images of the creatures, transparent relatives of jellyfish, the tubular constructions of their bodies laced with cilia that waved for propulsion and bioluminesced in the dark water.

Trapped heat in marine environments is dangerous in so many ways, but for a shallow marine ecosystem like the Gulf, it means an increased susceptibility to marine heat waves. Warmer water is also less able to hold oxygen than colder water, and oxygen depletion can have cascading effects on ecosystems, resulting, finally, in hypoxia or anoxia. Low-level impacts from marine deoxygenation includes increased susceptibility to disease in marine species, the reduction or increase of habitat ranges as species attempt to avoid the lower oxygen levels, problems, even crises, in reproduction, and finally and most terrifyingly, suffocation. The results can produce dead zones, areas where little to no ocean life remains. It can destroy the life cycle of phytoplankton, then by extension, zooplankton, and finally, the entire chain of animal life that depends on this complex marine food web.

The night we watched the documentary on dead zones, Dex woke up crying, "What if, after the nitrogen, all that is left is the comb jellies? The comb jellies are beautiful, but they can't be the only ones!"

Comb jellies occur all over the world. Technically very large plankton, they are extremely flexible feeders, eating other plankton, sometimes even one another. Recent scientific inquiry into their flexibility has shown that they may overproduce larvae of their own species in times of low food availability so that they can consume their young when the food is gone. Adaptability is their key strategy, ruthlessness, a mobility that makes even the worst of collapse a potential benefit to their survival.

Nitrogen runoff is a principal cause of algal blooms that produce red tides. Farming under industrial capitalism requires the false abundance of added fertilizer. Nitrogen fixers can't be absorbed by the soil in the same way that organic fertilizers can, and when they are at a saturation point, they simply wash away to the sea, creating those dead zones, where no life can grow except the heartiest animals. As with HABs, nitrogen runoff creates too much plankton, and that huge food source attracts the hearty, survivalist comb jellies. In the Black Sea, the comb jelly called the sea walnut, *Mnemiopsis leidyi*, arrived as an invasive species in ship's ballast, just as zebra mussels came to the Great Lakes and common periwinkles came to Maine. Without·natural predators, they crashed the ecosystem within a decade. The Black Sea and its fisheries were only saved by the accidental introduction of another species of comb jelly, the *Beroe ovum*, the sea walnut's natural predator. It was this crisis that led scientists to wonder how to stop a creature that can lay more than 12,000 eggs in two weeks, that produces young as a potential pantry for hard winters, that could create an ecosystem empty but for transparent death rainbows, a gelatinous sea filled with animals we have brought to the wrong places.

The chemicals in the dead zones in the ocean, the chemicals in my mother's body, they felt of a piece by the time I finished reading her books on climate collapse. I reached for empathy for her, for the things she'd given away to keep us alive. They were not different from the things I was giving away to save Dex. Maybe he would look at me at sixteen as I had looked at my mother, judging choices made in panic, under stress, not one of them humane or thought out. But the frenzied fear of loss said to her, *Work will save the family*, and said to me and maybe my father, *Refusing to die will save the family*. Neither choice was right. Both could result in beauty for a minute, shale dinosaur footprints in the sun, comb jellies lighting dark water.

By the river, looking at those footprints, Dex's hand in hers, my mother yelled, "Come on! Come down and see this!"

"Come on, Mom! Grandma found something wonderful!"

I saw them, their feet in the water of the Connecticut, leaving traces.

They walked onto the porch at the Gosnold Arms Inn just like that, holding hands. The set of their steps was the same, the way they observed the trees around them. We rested in the cool of the cottage room, everyone's shoes off.

Before dinner, we drove to the Rachel Carson Salt Pond Preserve. My mother ground the gears and lurched from the hotel parking lot, worried when she joined traffic, parked at the pond hurriedly and at an angle. Dex opened the door and ran to the steps that led to the beach of large stones and boulders that ringed the water. My mother and I took our time, walking slowly over the rocks. My vertigo was intense, my exhaustion too, but we stood together as Dex took off his shoes and walked to the water, a net in one hand, a bucket in the other.

"Just like you," she said, looking at him. And then she took her shoes off and walked into the water to join him. Bent down, looking for crabs and sea stars and lobster babies, they were as twinned as they had been at the dinosaur footprints, as twinned as we might have been when I was small, as twinned as I wished we'd been.

Once back at the hotel, she parked the car and we walked to the restaurant at the bottom of the hill, Shaw's Wharf. She beamed at everything, everyone. "I remember this place. I remember this smell."

She and Dex held hands while I walked behind them, slow and unsteady.

Over dinner, the harbor below us as we sat on the deck of the restaurant, I said, "I'm still feeling pretty bad. I'll need help tomorrow. Help to get us all to breakfast, that's included in the room, and help to get us to the morning boat. The only one I could get was two towns away."

"Well, sure, honey. But I'll need help too. With my bags, and with waking up. I'm not used to traveling, you know, and I have a lot of painkillers for the fibro."

"I'll help, Grandma!" Dex chimed in. "I can carry your bags."

"That's OK, kiddo," I said. "The grownups can manage."

Dex leaned into my mother on the bench. He had been smiling, but that drained from him once he was in her embrace, as if he was finally realizing how much he needed her. The wounded boredom of the hospital came back, the quiet. "I'm tired, Mom," he said.

"I bet, Monkey," I said. And he folded deeper into my mother's side.

Great White Shark

(*Carcharodon carcharias*)

The ferry ride to Monhegan was hard, the walk past the apple tree wrenching. My mother bounced along through all of it, ignoring how Dex shuttered in on himself when we walked past the bend in the road where I'd fallen.

We ran into Jen, who recognized me from the fall. "You're back!" she said and hugged me.

"We are . . ."

My mother talked over me. "My family lived in Maine," she said, "right on Southport Island! Elliot Winslow? Do you know him? He ran tugboats, and Robinson's Wharf, the lobster place."

"Oh! That's nice. I don't know him." Jen looked at me. "I'm glad you're OK. We can catch up later."

The hotel and the trucking company had helped to get our bags to the cottage. We walked slowly through the village to get there, past Swim Beach and the Fish House, past the grocery store and the hotel and the church. We turned up Horn's Hill and into the cottages perched there, and came to our cottage, tucked into the side of the hill overlooking the wetlands. A long, pitched, wooden walkway led from the dirt road to the front door of the cottage, from which you'd have a view of the harbor and the sunsets. Holdfast House was rickety and tilted. Its lines sloped. Once inside, we could see that the house turned in on itself and brooded. Mysterious water leaked from the light fixtures in the kitchen. I lay down on the tilted bed and tried to shake off the dizziness. In the next room, Mom went to bed. She didn't get up until the sun was setting, when Dex and I walked to the village for pizza to bring home.

"Grandma didn't want to walk with us?" he asked.

"She's tired," I said, trying to keep my voice neutral. I was tired. I'd hoped she'd help.

The next morning, she was still resting. "Dex wants to see the tide pools," I told her, hoping she'd volunteer. My dizziness competed with a worry that Dex's nerves, still frayed from the hospital, would snap in that place without routine.

"I'm so tired, honey," she said. "Maybe a walk after lunch."

I walked to the tide pools with him alone. At Lobster Cove, Dex played in the water, read his book, filled his bucket, talked with strangers, with a mania I'd never seen in him, a fury to put things all back in place as they had always been. I knew that mania in me. I remembered it from Dogshit Beach. I collected seaweed into piles with him, I lay back on the hot rocks and let my head spin, tried to check my pulse and guess at my blood pressure. We walked to town for groceries. I counted pennies and bought the cheapest things. At the house I made plain pasta with butter. After lunch we walked back into the village. This time Mom came. She shuffled slowly beside us, obviously in pain, trying to respond to Dex.

"That's Swim Beach!" He ran ahead of us both.

We followed him and watched families swimming in the cold water, whooping and joyful. Monhegan had been an artist colony for more than a hundred years, and on the rocks next to the beach, a painter was perched with a portable easel and canvas, a sun hat and backpack. My mother had begun painting watercolors in the years after the divorce, and she went to talk with the painter. When she came back, she said, "I'd love to come back here and paint."

At the toy store, I bought Dex a plastic gun that shot bullets made of bits of potato.

At the counter he asked, "Can I have a Tootsie Roll, Mom?"

"Yeah, Mom, can we?" my mother asked.

We walked back to Holdfast House. We made dinner and put Dex to bed. Sitting on the threadbare couch, I asked her: "Why would you say something like that? I'm not your mother. You're mine."

"Fine. I can't say anything right to you."

We sat on the couch in silence.

When Dex was six, as my divorce dragged on, after she'd moved away from us in Chicago, my mother rented a cottage on Cape Cod and asked us to stop there on our way to Maine. She wanted to do what Mac had done, providing a cottage on the beach for us in the summer. I debated going.

"Remember, this is your mother," Amanda said to me. "She'll ruin it given half a chance."

"Maybe," I said. "Maybe this is the year we figure it out."

She'd greeted me after twenty hours on the road by panicking that I'd parked my car wrong. Over the course of the five days, we fought constantly. We drove to Provincetown, parked in a huge, hot, crowded lot, and boarded a boat full of tourists going whale watching in the bright July sun. Dex wore a striped shirt. He held my mother's hand, skip-jumping up the walkway to the boat, his grin missing a tooth. They walked to the railing, my mother pushing past people until they stood together at the very bow, as close to the water as they could get, leaning out as if they could swim to the whales faster than the boat would go. I stood behind them, watching their bodies, the way genetics made them rhyme, the tilt of their heads, their slightly stooped shoulders. Dex turned and grabbed my hand with his free one, my mother and I linked by Dex, leaned into the spray as the boat motored out into the Atlantic.

"To port!" the captain said over the ship's loudspeakers. "We found the pod of right whales we see here sometimes!"

To port, close to where we stood at the bow, were the rising humps of slick black animals in the water. Sprays of aerated sea came up. The boat slowed, easing closer to them. We leaned farther, all of us, holding hands. And as the prow came as close as it was legally allowed to where they floated, the whales came up to breathe and we could smell the vegetal scent of the air that had just been in their lungs.

"Broccoli!" Dex yelled. "Her breath is like broccoli!"

On the boat, my mother went to sit by herself, exhausted from the energy she'd expended getting to the cottage, setting it up, paying for it, finding groceries and linens for Dex and me, booking the boat, picking out the beaches we'd visit, packing the folding chairs, trying to craft each moment as Mac might have done. She was getting sicker every day. I could see her through the milky windows of the cabin where she rested on a hard, plastic bench. Dex and I stood at the railing together, looking for more whales.

"A fin, Mom!" he yelled. "A fin! A shark fin! I saw it!"

"Folks, we're getting word that some of you saw a shark to the port, and that's pretty new for us here off Provincetown. Probably a great white that's been spotted here recently. I've only seen one of those before in all my years on this boat. If you get up and come to the windows or to the port side, you might still see him. But they move fast, so you could catch him off the starboard."

Through the windows I could see Mom looking around for us. She caught my eye and started to get up, scrambling for her hat. I looked back; the shark fin came alongside the boat, then ducked under it, headed to the other side. I pointed to the other side of the boat for her, and she turned her head.

"Oh! There he goes, folks! Now try looking for him on the starboard!"

My mother looked confused, slowed down, then raised her arms up in a gesture of defeat and sat back down again.

"Well, there he goes, folks! I hope some of you were lucky enough to see it. They may not be rare here for much longer, but right now, that's a new sight on the Cape!"

Sharks are seen more regularly now off the Cape. They have caused injuries and fatalities in recent years. They were spotted in the Long Island Sound for the first time ever in 2019. They are spotted in the Gulf of Maine. But their story isn't only one of predators moving north into warming waters, not only the result of collapse. In some places, the Gulf particularly, the return of great whites is as much due to marine management success stories as to the rapidly heating ecosystem. They come for the seals.

In 1972, Congress passed the Marine Mammals Protection Act (MMPA). This law overturned decades of common practice by fishermen, who hunted harbor seals in the Gulf and the Sound and throughout the Atlantic coast of the United States, capturing what were then one of their largest sources of competition for fish. The MMPA made it illegal to hunt seals. The unlikely friendship of a Maine fisherman and a seal called André who lived just up the coast from Monhegan made seals into loveable, anthropomorphic clowns in the popular imagination. Between the image makeover of seals and the new legislation, the seal population bloomed. The sharks came back for their favorite food.

In 1976, Congress passed the Magnuson-Stevens Fishery Conservation and Management Act, which was intended to create a more conservation-focused plan to manage American fisheries. In 1997, an amendment passed making it illegal to harvest sharks in Atlantic federal waters, protecting the great white under Magnuson-Stevens. Great whites are a sign of transformation, the return of an animal that is "newly" observed in some waters that has belonged there all along. The University of New Hampshire's Sea Grant information page on the great white puts it this way: "As conservation and management efforts in the Gulf of Maine ecosystem restore populations of marine life, we expect to see examples of species interactions that appear to be "new" to us given our frames of reference and ways of living."

It is never as simple as a single story in the life of an ecosystem or a family. There are second chances and failures all in a single day, a single week.

"I didn't see him!" Mom said to Dex as he ran up to her on the whale watch boat.

"He was amazing!" he yelled, and threw himself onto the bench next to her, butted his head into her arm. "Amazing."

"I wish I'd seen him." She was starting to cry, as if all the work she'd done to make a vacation for us had still deprived her of the thing she wanted. As if she hadn't stood with us, five feet from the plankton-broccoli breath of whales.

"I'm sorry, Mom," I said, "maybe next time."

"You know we won't have a next time."

"Next time, Grandma."

As we sat on the couch on the ground floor of Holdfast House, looking at each other in silence, I realized this was our next time. We had the chance to revisit, even revise, that trip to the Cape, to revisit The Periwinkle, to revisit Mac and Dad and every time we'd come to the sea together to face what was hard. I was sick, she was sick, there was a child who needed parenting, care, security, wonder, joy. There were two of us who could do it. She had the chance, ill or not, to stop needing me to mother her. She had the chance to lean into the power she had, to face the truth of my vulnerability and be as strong and as soft as she always was. I remembered what I said to Amanda before the trip to the Cape—*Maybe this is the year we figure it out.*

"I need you to be the mom for a while, Mom," I said. "I can't take care of you and Dex at the same time. I'm getting sick."

"I can't, sweetie," she said. "I don't have anything left."

Eastern Oyster

(*Crassostrea virginica*)

I tried to sleep in the weird bed I shared with Dex in the Holdfast House, my mother's words in my head. Sometimes it is harder to love the world that is both broken and whole. Sometimes it is harder to love the people who are both strong and weak. Sometimes it is harder to stay present to the truth, that nothing is going to return to full wholeness after loss, but that you might still find a way forward.

The floors of the upstairs of Holdfast House were slanted, and the bed tilted. I made myself the barrier to Dex falling out in the night. His body was rolled against mine when he slept. I caught myself from falling out often as one leg started the drop to the floor. Rest was hard. In the morning, my head spun. I propped Dex up with pillows and walked carefully down the slanted stairs to the kitchen. I was losing weight, unable to eat, exhausted, indistinct in what I heard and remembered. Rage thumped at the back of my throat. *She had nothing left. She said so.* I had nothing left. I began making breakfast. She came downstairs.

"Can I help?"

"Make toast?"

"Sure."

She began humming the lullaby she used to sing to me when I still couldn't sleep after our Flopsy Flora game, the lullaby about the girl who died of a fever when no one could save her, *alive, alive-o, crying cockles and mussels alive, alive-o.*

A fierce wave of wanting fell over me. I wanted this to be a normal vacation. She had been the one to say it, but I wanted that too. I wanted to be the kind of family that took a cottage and spent a week walking around the island and the village. I saw them all the time. They sat at The Novelty, eating ice cream cones and laughing in the late sunset, walking down to Fish Beach to see the red ball dip into clouds behind Manana. They sat on cliffs with sketch pads or sandwiches. They shopped for paintings, and they went out on sailboats. After all this time, I just wanted that, a simple thing I told myself I'd never have the summer I was diagnosed, a simple thing I'd tried to find in the safe, quick marriage to Jonah, a simple thing I'd been back and forth about deserving for more than thirty years, maybe even since that summer I was nine and tried to understand desire and loss all at once before I ever had language for it. All I ever wanted was time with the people I loved. I could not name it as desire. I could not name desire at all.

The moments with my mother at the Sound, the way she told me about how PKD had taken sex from her and my father had made me into a deeply private person, unwilling to talk about sex, unwilling to write about it, unwilling to admit desire in person, even to the people I dated and married. Desire sat behind a wall that I had begun building out of shells and bones, sand and fish, all on that single rainy night on Dogshit Beach when I saw who all of us were, my family, trapped together in the mouth of collapse, brittle and solitary and sad. But, in some strange contradiction, though I wouldn't talk about sex, I pursued it with fierceness, pushing my body to be a vehicle of joy apart from its pathology.

Standing in the kitchen of the crookedest house on Monhegan, watching my mother make me toast, my body was telling me that I couldn't breathe if I kept living that way. I couldn't build a future until I let desire for life move through me unfettered, just like anyone else.

Since my diagnosis, I had tried to learn how to live. I had tried to remember that the whirlpool was also the vortex of the stars. I had tried to open the doors I saw well people open. But beyond each open door there was a room without a floor that emptied straight into the dark water of the whirlpool. Dog-paddling in the cold black, I held myself where I could breathe, a knife in my teeth, waiting for anyone to come into the depths and tell me I did not deserve to live. I watched others in the family drop their knives and let the whirlpool take them. But I held mine. I listened to the doctor tell me, *I know how to end PKD in one generation.* And the anger got fiercer, the separation from the well more profound. Sometimes I climbed out of the whirlpool to pretend I was like them. But then I'd open a door and it would begin again.

Oysters are freighted with moral judgments—the lure of them, the sex, the desire. I know I have desires, desire for the sea, desire that brought me to marry and marry again, like grains of sand inside an oyster, a secret, an irritant. I want to be seen and loved like Lois was. I want to draw in against the sea, close my shell, and hold the irritant close, so it will always make a pearl.

Profligate, oysters renew and defy the fishery. In the oyster farms that fishermen turn to as their old fisheries collapse, they grow the creatures in mesh bags. In recent years, the wild population of oysters has exploded. Scientists wondered if this was from the warming Gulf, then realized that the captive oysters were breeding from their mesh prisons, sending seeds, called *spat*, into the open sea.

The first oysters I ever tried were those Jonah ordered on a date with me at a restaurant we couldn't afford. Eating them was like drinking the ocean. We went back to the same restaurant after we were engaged and ate them under a sign on the wall quoting Hemingway, "As I ate the oysters with their strong taste of the sea and their faint metallic taste that the cold white wine washed away, leaving only the sea taste and the succulent texture, and as I drank their cold liquid from each shell and washed it down with the crisp taste of the wine, I lost the empty feeling and began to be happy and to make plans." We put the quote inside our wedding rings, *to be happy and to make plans*. My father yearned for them at every trip to the sea. My mother refused to stop in the Wellfleet oyster shacks we passed on the side of the road. In 1980, my father sat with his former apprentice, Dan, in Dan's garage on the coast of Maryland, between them a bushel basket full of Chesapeake oysters.

"I can't believe they let you buy this right off the boat," my father said, then shucked an oyster straight into his mouth. I stood watching them, leaning against the frame of the open garage door.

Dan laughed. "If you lived down here, we'd do this twice a week!" a beer was on the concrete floor next to his feet, a beer on the cement floor next to my father's.

"If I did this twice a week, I'd be dead in a month!" he laughed back. "These are terrible for my kidneys!"

There are medical and cultural opinions about oysters' ability to increase testosterone in men and women, but also articles on the rabid sexuality of oysters themselves, their former abundance, the history of oystering in the Sound, the work of transitioning lobster fishermen in the Gulf of Maine to farmers of oysters, places where oysters are being reseeded in beds to prevent the coastal ravages of warming, rising seas. They are caught up with all these desires—for work, for survival, for permission, for connection.

Somewhere oysters are mating, clouds of sperm in the ocean, clouds of eggs. Oysters mate abundantly, extravagantly, each female producing fifty to a hundred million eggs per spawning event, a shower of offspring to make up for the cruelties of the ocean, the life of their larvae the life of all mesozooplankton, adrift and consumed. When oysters mature and settle, they prefer the comfort of other oyster bodies, layering in reefs, mounding up in stacks of confederates until they form barriers that protect the coastlines. But then we take them from one another. Our desire has transferred from our own bodies to theirs, ravenous, consuming, destructive.

My father's desire once must have been like that, for oysters, for women, for my mother in her beauty, the brilliance she couldn't see but he could. He tried to get her back until he died. She visited him in hospice, and they held hands.

I wanted to be loved, when I saw that.

Six months after my father died, I met Jonah. We met at Amanda's wedding in June, and though he was a work friend of her groom, unlike the artists I'd dated, there was chemistry. I was so flush with mourning that his optimism felt like joy. We wrote email love letters to each other while I was in Massachusetts settling my father's affairs. I left my mother's dusty basement full of his tools and went to an internet café and wrote long paragraphs about tomatoes warm from the garden, and he wrote back about the Chicago summer I was missing. By the time I realized how much our differences strained us, the World Trade Center came down, and I mourned for the city I was born in and retreated deeper into the relationship.

In October of that year, we took a vacation to Door County, Wisconsin, on the shores of Lake Michigan. Over lunch in a restaurant famous for the goats on its roof, I asked him what he wanted.

"I want to do this," he said, "make something of this, of us."

"I am going to be sick someday," I said. "I might not be able to have children. I might be sick soon, you know. Are you sure you want to do this?"

"Yes."

"I mean, *sure* sure? Because everything I have to deal with is out of proportion to normal life, you know?"

"I think you're just life-sized," he told me, and a warm flood of gold poured through my chest. No one had ever said anything to me that mattered more. If he had asked me to marry him right then, I would have done it.

Instead, we drove from the restaurant and spent two days in a tiny cottage called Tamarack that was suspended slightly over the shore of Lake Michigan, its 1970s paneling and huge picture window outdated, but its bed pushed up so close to the view that when we woke, naked and hungry, the moon shone in a line down my back.

A few months later, I went home for Christmas with my mother and told her the story of being life-sized. She was thoughtful. "Do you love him like Lawrence?" she asked.

"I won't love anybody like that."

"Do you love him enough?"

"He understands the curse, that's a good start."

"And he has a job? Insurance? No dead mother?"

"No dead mother."

"Then take something back with you," she said, and went into her room and brought out Mac's engagement ring, a 1930s diamond Doug had given her.

Jonah picked me up from the airport, and next to the ancient gas heater in my cheap apartment he blurted out a proposal. The ring was in my pocket.

Then we were eating oysters and being happy and making plans and then we were married. I hadn't been wanted by Lawrence enough to be chosen, and Jonah's desire, his confidence, his insistence, undid all the hesitation I'd had with Ben, all the loss I'd felt with Lawrence. Jonah made me believe that we could have a life, that the disease didn't matter. I didn't wonder until years later if he truly understood what it meant to marry into PKD. I didn't know that there was another way to be loved, one that really did understand what I was fighting, that saw me as desirable anyway, that chose me like I chose to have Dex.

All I knew was that I wanted to believe in a future. Surviving means you don't drown in that whirlpool. Surviving means that when you hit the black water, you find a way to swim through the depths.

Desire for a future meant that thousands of people were once employed in the Sound oystering. Oyster shacks covered the mouth of the Housatonic and the Connecticut. Desire meant that as the waters sickened, the oysters did too. As Andersen points out in *This Fine Piece of Water*, in 1892 Wesleyan students at a frat party sickened from raw oysters they'd eaten. In 1924, 1,500 people on the East Coast and in Chicago contracted typhoid fever from the oysters of New Haven. When 150 of them died, it nearly took the industry with them. But desire returns and returns. Until multinucleated sphere unknown disease (MSX), a protozoan infection lethal to eastern oysters but not humans, threatened the oyster fishery, Connecticut had robust oyster harvests in the twentieth century. As that disease has been limited and the state has created programs of public oyster beds and encouraged oyster harvesting, the fishery has rebounded. According to the Associated Press in 2021, "Government and private efforts to clean up Long Island Sound, develop disease-resistant shellfish and find new markets have helped aquaculture become a $30 million a year business in Connecticut, with oysters accounting for about half of that." In 2022, there were more than $8 million in oyster landings in Maine, where twenty years previously, landings accounted for less than $700,000. This farming is sustainable, profitable, manageable, a way to earn a living from the sea, commercial oysters still, even in farms, filtering nitrogen from the water.

The only desire I admitted was for the other things a body could do while it was well. I desired walking in the woods, driving twenty hours through the desert, finding phosphorescence in the sea. I desired music, which was every desire I'd ever had unlocked in a single practice. On stage I felt a column of light move through me, something larger than sex, something sex contained in the personal opened up to become the universal, a thrumming wonder, a beating heart, the will to reach for all the wanting.

In bestiaries, chimeras are a symbol of nature broken and deviant, and sometimes a symbol of female evil. The name is based on the Greek word for *she-goat*, with the head of a lion, the body of a goat, the tail of a snake. Genetic chimeras exist in the plant and animal worlds: bicolored flowers, intersex humans, merged twins, these are all chimeras resulting from the merging of distinct zygotes—all miracles, all wonders. A friend of mine with PKD has become a chimera. Her body is her own, but her kidney belongs to another person, another gender, another life and genetics. Before she was gifted that kidney, her old kidneys were removed, and she was also given a bone marrow transplant from the same man. Before that gift, her old marrow, her old immune system, was destroyed. The new bone marrow was transplanted, and then the new kidney added, so that, in a series of steps taking months, she became a hybrid, key parts of her borrowed from another human. Her new immune system was called a chimeric immune system. Her chimeric immune system meant that she had to be vaccinated again for all the things she'd been vaccinated for in childhood, her immunity erased. It meant that she had a new blood type. Her new kidney now recognized her, since it came from the same body as her new marrow. And the organ joined her body in concert, meaning no immunosuppressant drugs were required for her to live, a cozy chimera, a happy hybrid.

"And you don't have to take anything? Nothing at all?"

"No. I mean, I lost a year to this. But no. You know, I could give you the name of the specialist who did this. I was in a trial, but they just got funding to expand. You may not be eligible, you know, there's a battery of tests. But you might. You don't know until you try."

"Yeah, yeah. I'd like his name."

"Oh, Eiren, why are you crying?"

"I'm thinking about my dad. Immunosuppressants basically took his life in the end. I mean, it took ten years. But if he hadn't had to take them . . ."

". . . it would have been a miracle."

"Yeah. Yeah."

"Maybe that will happen for you."

"It is just so incredible."

"I know. I know."

Desire for a future where I find healing flooded through me when I sat with her, crying over miracles. In her family the illness was a spontaneous mutation. Her mother died of a brain aneurysm before she even knew her kidneys were compromised. When my friend got her diagnosis, her nephrologist said, "Yes, this disease is terrible, but you must remember to dance!"

Desire for a future for me and for Dex floods me still. I wait to see what will come. Immunosuppressants and fistulas and transplants and virus screening and dialysis and chimeric immune systems and the first drug prescribed to slow the growth of cysts have all come since my father was born. I cannot picture the wildness of the world that will follow me, that will follow Dex. Each time, the wonder that allows us to live longer and longer, with more and more dignity, with more and more hope, leaves me as breathless as I was the first time that I felt the blood moving through the fistula under my father's skin—thrilling.

I think that this is called *desiring my own life*, believing in my own future. I think that no one else can give it to me. I have to believe in the necessity of my own body, my own spirit, my own future and path forward. I think I am constantly relearning this belief. I think we are constantly relearning to believe in the world this way, and to believe in all its creatures.

We want the restoration of the oyster. And in that wanting we propose legislation. The Living Shorelines Act, proposed but likely dead, would fund NOAA to support communities seeking solutions to coastal erosion and habitat loss by, among other things, restoring oyster reefs. In New York Harbor post–Superstorm Sandy the coast is being seeded with oyster shells collected from restaurants through the Billion Oyster Project. The shells are taken to Governors Island—by 2020 more than 1.6 million pounds— and then cured in piles like middens and returned to the ocean as the foundation for oyster larvae to create new reefs. As of 2022, more than 100 million oysters were introduced back into New York Harbor since the project began in 2014. The goal is for that project to create one hundred acres of new oyster habitat by 2035.

But those numbers, those advances, feel like the stolen moments of desire I remember—the hour onstage to come alive to all the wanting when the twenty-three hours surrounding it were silent and contained, a day permitting rage in a year of swallowing it. In 1609, when Henry Hudson sailed into New York Harbor, there were an estimated 220,000 acres of oysters there, perhaps half the world's oyster population. They are nearly gone now. Worldwide, the number of oysters in estuaries has declined by 80 percent since white settlement of the Americas. It is hard not to feel rage at a goal of restoring a hundred acres where there were 220,000.

I feel anger at the base of my spine as I think of the loss—oysters and chances and joy and desire. *I know how to end PKD in one generation.* The anger fills the bowl of my hips like an ocean rising, cradling my kidneys, so outside their normal confines, and moving up to burn the things I want to say. It is not enough—to take the pittance of desire on the stage, to take the pittance of *to be happy and to make plans.* One hundred acres of oysters is not enough. Fifty years of life is not enough. There should be joy and desire and sex and wanting and health and oysters filling the ocean with seed, like a phosphorescent trail behind a night swimmer.

My mother and I should have been able to find joy in that crooked house on the island that was my favorite place on Earth. But as I worked and she sang, the fuzzy hum of my brain keeping me from remembering how to clean up after breakfast, what recycling material was supposed to go in what bag, she leaned in to help me.

"There's just so much to do," she said flatly, exhaustion in her voice. "You'd think they'd make a vacation a little easier."

"This isn't a vacation, Mom! This is a problem! I was in the hospital three days ago." She looked like I'd struck her. "I don't even know what's wrong with me, you know, or if I'm going to have another seizure. I didn't ask you here to vacation. I'm supposed to be working, that's why I came; you haven't even asked if I need anything."

She hung her head. "I'm sorry I'm such a disappointment."

I sat down in a chair in the kitchen and stared at her. I wanted to be furious, to blame her not only for the crooked house and the seizure, but for everything, for not protecting me from the disease I carried. She had been fragile and brittle, vulnerable and needy. I wanted to be vulnerable, to desire healing and care, to want things for myself. It all crashed down on me, the shabby kitchen, the simple breakfast, the island I wanted to spend more time on, the parents I'd wanted to have, the future where I didn't have the disease at all.

But the disease wasn't her fault. I could see her failures, but for the first time, she seemed just delicate, not something I had to keep from breaking, or something that could keep me from breaking, but a shell over tenderness, a jingle shell made just the way it was meant to be. "I'm sorry," I said. "You aren't a disappointment. This is all just very hard."

"I know, Monkey. It's been hard for a long time. I'm sorry."

A rush of calm moved through the building, a breath. She reached out to hold my hand. "We'll figure it out."

"The recycling too?"

"That too."

She stood up and walked back to the kitchen and began to wash out plastic tubs; she sang, *She died of a fever/and no one could save her/and that was the end of sweet Molly Malone.*

I stood up and started making toast for Dex in the narrow kitchen, the screen door open to the smell of the harbor. *Now her ghost wheels the barrow/through streets broad and narrow/crying cockles and mussels alive, alive-o.*

North Atlantic Right Whale

(*Eubalaena glacialis*)

"Let's go to Fish Beach," Dex said when we came back in from getting pizza down the hill later that night. My mother was on the couch, and we took the box to her, put it on the coffee table, and began to choose slices. "The sun is going to set in a little while. And it is our last night on Monhegan."

"That's right," she said. "I can't believe it's the last night."

"I don't know if I can walk all the way down there after getting dinner," I said.

"Just see."

"Yeah, Mom, just see!"

"And we can get ice cream."

"Ice cream!" Dex said. "And maybe we'll see whales! Like we said, Grandma, *next time*."

Right whales form a matrilineal culture—*culture* meaning shared information, tradition, routes, languages passed down from one generation to the next. Right whales are large, fifty feet long on average, long-lived, females reaching seventy years old. There is speculation they could reach up to one hundred if unimpeded. But they have long gestation periods—thirteen months—and the deep bond between mother and calf ensures species survival but makes rebuilding a robust population difficult. Calves stay with their mothers for more than a year, learning the culture of their community. There are photographs of them cuddling in the sea, the mother wrapping her flippers around her calf. The matrilineal bond is so strong in some baleen whales that early whalemen would harpoon a calf, hold her, injured but alive, until the females of her group would gather, then kill mothers and grandmothers alike.

Mothers whisper to their calves while traveling through dangerous waters, calming their children while sharks or predators are close, so that only the babies hear.

As the Gulf warms in its complexity, commercial fishing patterns change to respond to the chaos. Fishermen, already under economic collapse, jettison their extremely valuable gear to comply with calls for whale-safe fishing lines to prevent entanglement, new routes to prevent boat strikes. The costs of these changes fall entirely on those working-class families, when technology already exists to respond to these crises with support from the state.

Philip Hamilton, North Atlantic right whale researcher, has worked with right whales since the 1980s. His superpower is identifying individuals in the once massive groups by the distinctive patches of white on their heads, callosities, appearing in patterns unique to each individual, home of white sea lice. Once numbering 21,000 individuals in the Atlantic Ocean, *Eubalaena glacialis* now number fewer than 365. In 2020, I called Hamilton in Lubec, Maine, his home base for studying the right whales of the Gulf and the Bay of Fundy. I asked him how he felt about the losses in those communities over his career.

"I've stopped using the phrase *Something bad happened to me.* I've begun to use the phrase *Something powerful happened to me,*" he said.

"Really?" I asked him.

"Yes. I find that that is potentially one of the benefits of the right whale population tanking so quickly."

The day was placid, but my chest was suddenly ringing. An interview subject had never kicked the wind out of me like that with pragmatism born of bearing witness to extinction. I had a hard time seeing my illness as powerful, seeing the whales in the Gulf dying as powerful. Hamilton watched all those families die, but was transformed.

Right whales can eat between five and six tons of plankton per day. Their migrations from the southern reaches of the Atlantic coast to Canada are timed to coincide with the usual emergence of zooplankton from winter hibernation. When they arrive ahead of or behind the plankton arrival, they go hungry, and so do their calves. In order to combat the scarcity, right whales have to improvise, change up their feeding patterns, not easy to do on the fly for a species that relies on generational culture.

They may have to dive deep to find plankton closer to the sea floor. While diving for prey, they are far more likely to encounter fishing gear, especially traps and lines used by lobstermen. More than 82 percent of right whales have scars from entanglements with fishing lines, with some individuals showing signs of being entangled more than seven times. The Maine portion of the Gulf of Maine, in which one finds the typical foraging grounds of the right whale, has more than 900,000 vertical fishing lines. When the plankton are at the surface, they may have to rise to the warmer water at the top of the water column to reach them, and there they are far more likely to encounter the shipping lanes, pleasure craft, and fishing boats, and die by boat strike. Conservation organizations have been working for decades to protect the last of the planet's North Atlantic right whales, and much of the policy and legislation they have proposed has been adopted by the fishermen

and lobstermen of the Gulf of Maine, whose conservation practices have been scrupulous in terms of understanding and adjusting
to the whole ecosystem. But as the Gulf warms, as the Atlantic
warms, and as animals move north and disrupt their predictable
patterns, those practices need to adjust further.

What brings change can be complicated. By the time she was
diagnosed with colon cancer in 2017, my mother had been sick for
a long time, praying for compassion mind, seventy-three years old,
saying mantras. The chemo disoriented her. Her personality slipped
further into childhood, into dependency. I flew back to help, but
she kept falling and hurting herself alone in her apartment. Her
care team sent her into a nursing home where her state changed
again, to one of rage and forgetfulness, defiance, confusion. They
called to tell me to come home and sort out her last weeks.

Andy, my second husband, and Dex and I were on a trip out
West when the call came, taking me to a residency. Andy, the poet
of labor unions and glaciers, Andy, the man who had been a boy
reading *The Mabinogion* in a cheap rented house in Georgia when
I was a girl reading *The Mabinogion* in a cheap gas station apartment in Massachusetts. Andy, who loved and understood Dex
from the moment they met. Andy, who hiked with us and fell in
love with Monhegan and took us to forests we'd never seen in our
own backyard. Andy, who proposed to me on impulse, standing
in a Wisconsin basilica named for Mary Help of Christians, where
pilgrims flock for healing and discard hundreds of crutches. Andy,
who stood with me, engaged for only three hours, in a forest with
wild deer while the stars came out overhead. Andy, who I married
in 2017, two weeks before my mother knew she had colon cancer,
when Dex was twelve, standing on a glacial kame in a forest preserve two hours northwest of Chicago and looking down at a creek
filled with freshwater mussels.

There were once eighty species of mussels in Illinois. Only fifty-nine of those species have been seen since 1970. The mussels in the Nippersink Creek include the giant floater mussel, which is so large it fills up two outstretched hands. We said our vows before Amanda and before Dex, who watched until we asked him in and vowed things to him and sealed those vows with a trilobite fossil pendant that his new father fastened at the back of Dex's neck. He has never taken it off.

Before the call from my mother's nurse, we drove west from Chicago through a catastrophic heat wave, into North Dakota and down, at last, to the Badlands. I found the back roads that led to Sage Creek campground, where Heather and I had slept surrounded by grasshoppers and bison and I had seen what the whirlpool could be if I could hold death and life together in the same body.

We parked the car and stood in the silence together. The bison herd was on the far horizon. "This place," Andy said. "This place is amazing."

I felt it filling me up, Dex running to the far ridge to see the herd, the same earth under my feet, the same pulse coming up through my shoes, and the column of light flowed up through me. We drove away as the sun went down and the stars came out overhead.

The nurse reached me as we were starting a hike in the Hoh Rainforest on the Olympic Peninsula in Washington State, on the edge of the sea I never understood, where orcas and seals were only a few miles from the forest of dripping, ropy moss. "She's not going to make it to the end of the month," the nurse said. "And she may need a hospice that will take her insurance."

I flew back. They moved her to a hospice I found for her, one with a harpist and a choir, and we waited for Andy and Dex and her sister, June, to arrive. I tried to remember lyrics to the Clancy Brothers' songs she'd played for me on vinyl when I was small, sang them until I trailed off. *The whistling gypsy came over the hill/ down to the valley so shady,* I sang, thinking of my father, the whistler, the dreamer. She slept. I sang the lullabies she'd sung for me. *She died of a fever/and no one could save her/crying cockles and mussels, alive, alive-oh.*

She woke, moaning, "We have to make the train. We'll never get to Rome like this." She slept. I slept, shoes on, my half-eaten container of applesauce on the floor next to me.

In the dawn light, when they woke her with more meds, she looked at me with suspicion. "You're trying to fucking poison me. You're trying to fucking take my house. You've been this way your whole life. You always hated me. You were always a terrible person."

Terminal agitation—sometimes called terminal delirium or restlessness—brings rage, hallucinations, belief on the part of the dying that they will be harmed, killed, detachment from the careful end of life plans already made. It may be the last stage a family sees before the dying leave them. Terminal agitation meant that my mother's voice in the dawn was the same as hers in 1987—me at sixteen, her at forty-three and drunk— shouting in the house we rented in the woods: "You were always like this, you lying little bitch! You never told me that you needed a ride to play rehearsal tomorrow. I won't fucking give you one, because you are a fucking manipulator and I fucking know what I heard."

All the poisons emerged from her mind in the hospice, the thought equivalent of the chemicals from paper mill runoff, the legacies of broken industrial New England, the stress of marrying into generations of curse, fighting the powerful stories polluters tell to keep polluting. The darkness dragged her down in her last days, telling her that all the evil was real and more powerful than she was and infesting everyone around her like leachate from a buried fuel container.

The hospice spiritual advisor found me in the corridor on the way to cry in the shower. "That's not her, that's not how she feels about you, you're doing a good job."

"I don't know."

"Are you angry? You have a right to be angry."

I was furious. Jonah, at first generous about the time Dex was away to see his dying grandmother, became impatient, texting that my vacation time had run out, that I should send Dex home, stabbing at a wound until it bled. Andy, worried about time from work and money, hedged about flying to the hospice. The insurance company had denied the payment for the nursing home. I was furious at my father for dying. I was furious at my mother all over again, as if we hadn't tended the wounds. I was furious at her for leaving me alone with my illness. I had tried to let go any hope that she would care for me. I had leaned into what felt like studied resignation that I would always have to care for her. I thought we had come to some kind of peace together, where I forgave her, or tried to, for all the ways we'd failed each other over the hard, ragged years. I thought the balm of my marriage to Andy, of my life with Dex, meant that the worst was gone. I had a family who cared for me, a community of friends who raised money to pay for all the expenses of the accident on the island, who came to see me marry Andy. We were healed, I thought.

The counselor was right. There was rage, rage behind every glass of water I brought my mother, every kind look. There were lakes of anger. "I don't know," I said to the counselor. "I don't know if I'm angry."

"That'll come back to get you," she warned me. "The more you feel your real feelings now, even anger at her, the better off you'll be this winter."

The willful girl I'd been in 1977—my mother thirty-three, me six, powerless, absorbing the chaos of that house—that was my voice as my mother died, *No, you can't make me.* I tried to remember the letter she'd written me after Lois died. The morning after Elvin confronted her in New York, I could hear them up talking on the couch in Lois and Elvin's living room.

"I know, Elvin," she said. "You're right. I'll lose her."

"You will. If you haven't already."

"I don't know how to stop."

"You just have to. You're strong enough, Nancy. I've seen what you can do."

At home a week later, she sat down next to me on my bed. "I'm not going to drink anymore, Monkey. I'm done. I can't do that to you."

"OK."

"And you can hold me to it."

"Whatever, Mom. OK."

"OK."

One night she slipped, drank and passed out, and when I woke her up to take her to bed, she said, "I broke it, I broke it."

I know I must have felt rage, betrayal. But all I met my mother's words with that night was silence, the freeze-out I'd learned was the only way to survive.

The next morning, she slipped a letter under my door. "This will never, ever happen to you again," she wrote, "and I promise I'll spend the rest of my life trying to earn back your trust."

She had spent her life trying. She never drank again. I had spent mine resisting. I didn't know how to forgive her if the round of healing and wounding never stopped in our lives together. I didn't know how to rest in the uncertainty of it and love her anyway.

Her sister, June, arrived at the hospice, and Tommy came from Brooklyn and things began to change. She was asleep more often. When June held her hand, she looked beatific. When Tommy told her to take her medicine, she complied. I went back to her apartment to sleep. She asked for fresh peaches from the farm stand, and I brought them and watched her eat the summer juice from them. Andy and Dex arrived the last day she was conscious.

"Finally, my boys are here," she said to Andy.

"I love you. I love you. I love you," she said to Dex, holding his hand.

"I love you, too, Grandma," he told her.

Andy kept my sleeping mother company in the hospice room when the rest of us took apart her apartment. The craniosacral therapist came. She told Andy that when she touched my mother, she had a vision of her with her ancestors in an orchard full of peach trees. "She was walking through, picking peaches," the therapist said, "throwing the ones away that were unripe, the ones that were rotten, then eating the good ones, while she decided if the snake wrapped around her should stay or go."

She died three days later.

My years of swallowing rage didn't work. I was unready when my mother died screaming at me, at the world collapsing. I couldn't scream back. I couldn't scream alongside. I lost her. Her voice is with me still, agitated, terminally. I don't want her rage to erase the peace she found in prayers for compassion as she learned to paint watercolors and be a good Buddhist. I don't want my rage to swallow me when I remember that my own mother called me craven, scheming, vicious, while I bathed her feet and sang to her.

After all, she was right. I was craven not to tell her I was angry. I was wrong to leave her alone.

I expected to hear anger in Philip Hamilton's voice when he talked about watching the whales disappear. I expected him to talk only about solutions, things we had to beg and fight for. Regulations have been proposed that would change, again, the types of ropes and traps used by lobstermen in the areas most likely to see right whale entanglements. This would mean replacing existing gear with lighter ropes that are easier for whales to escape from on their own, as well as lighter traps. For lobstermen, with a huge investment in gear representing most of their sunk costs, this could mean financial devastation. In Maine, this has lobstermen angry, and they are pushing back. The fishery in Canada has not been as responsive on the issues of either entanglements or boat strike. The period between 2017 and 2023 was so devastating to the right whale population that it has been designated an "unusual mortality event" encompassing injured, sick, and dead whales, which made up about 20 percent of the population.

In Massachusetts, adjustments made to boat speed regulations have been changed in response to increased boat strikes, but the habits of the whales changed again after those adjustments, and new strikes occurred outside expected feeding periods for the whales. The thinking among the lobstermen of Maine is that an overcorrection—replacing their entire stock of gear at crippling expense in a time when all sources are predicting continued decline in their fishery—is being requested in place of corrections that should be happening in other parts of the Atlantic. Evidence seems to show that even those adjustments may not work as the right whales' erratic behavior aligns with the newly erratic lives of the krill they eat.

Every effort at change moves them closer to safety, as their fate is bound up with the fate of everything on a heating globe. If the lobstermen of the Gulf of Maine don't shoulder the work of adjusting fishing and sailing practices alone, if they are part of a collective stretching from Georgia to Canada that alters structures, gear, and habits to protect whales, they might feel less rage. If conservation and regulatory efforts acknowledge the complex factors at work and coordinate, then the scientists are not alone with data and no agency. By using models that could predict the emergence of zooplankton, we might predict how best to support whales seeking their food source and adjust human activity. The larger question, the general one, the cultural one, of how to shift toward a planet repairing collapse, is the collective next step, one that insists that we learn and adjust together.

In 2018, Andy and I had a wedding celebration with friends to cement vows we made at the glacial kame. We did it quickly so that my mother, newly diagnosed with her cancer, would be with us. She put the chemo on pause and flew out in a blizzard and danced with our friends and new family. We told everyone that our marriage was crowned in the stars we'd seen on the stained-glass windows of Mary Help of Christians, and in the sky above the trail the night Andy proposed. By the end of the weekend, the snow had cleared and the airports reopened and my mother could fly home to the Sound's watershed and her apartment a mile from the Connecticut River. We drove her to the airport. Andy double-parked in the drop-off lane, and I unloaded her bags and walked her to the sliding doors where a skycap took her luggage and called a wheelchair for her.

"I'll see you in April, well, late March, only a few weeks," I said.

"Yeah, I'll see you then. We'll get you mussels at Joe's"

"We'll go to Joe's."

I hugged her close. Her body had become frail, I could feel her bones. Her short hair curled against my cheek.

All of our history seemed to braid around us, a net—part snare, part shelter. I assumed the disease was why she left my father, rejecting us both, and Dex too, by extension. I formed a list of animals as protection against anger, against missing her, against being left. I made myself a mother. I made myself a science writer. I made myself free. My father and I had the same disease, the same last name, the same unruly ability to risk everything for a cause, an idea, the same silences, the same habit of letting money run through our fingers. And if she'd left him, she'd leave me, so I left her first. I didn't realize that she was fighting her own mental illness, her own unique brain wiring. I didn't forgive her for that. But as we hugged, I understood that I was trying to heal all of it, doing the work of my spirit. When she was there in my arms, at the doors to the airport, I finally understood that I would have to carry her with me until I finished that work.

"I love you," she said. I could feel her start to cry. "I'm so glad you found Andy. I'm so glad you have Dex. I'm so glad I saw everything you made here. You made a good life here."

"You made that happen, Mom. You saved me and Dex. And Dad. You gave up your house. You worked so hard. You went to school. I can never thank you enough. I just want you to get well. You'll get to be here more often now," I said, hoping for more time, "and get to know Andy."

"Even if I don't, I'm so proud of you. You built it all back."

"I did. I tried really hard." I stared crying. She held me closer.

"Miss"—a security guard tapped my shoulder—"you have to move your car."

"In a minute." I looked at her.

She saw something in my eyes, and my mother's face, and took a step back. "OK," she said, "in your own time."

"I'll be out in March, Mom." I embraced her again. We were both sobbing now. I knew, I knew I'd never see her again as she was in that moment. I would never feel like I did. I would never feel the healing of the wedding, the toasts. We all took a leap of faith together and found peace and hope in that room, reading poems and singing songs to one another as the snow fell on Chicago and the planes stopped and the winter made the city silver and white and quiet. All of us were in that warm, yellow light in the loft, where art and music and bohemia and community had been made for twenty years. We took the lives my parents had chosen and turned them from chaos to safety. We said the names of the dead and ate food we made for one another and sang again until the last of us went home in the dark full of what can be made when we all love one another and lean into the possibility that life will keep on, and the dead will come to see it, and the stars will turn over a snowbound city and the lake will freeze and smoke in the cold, and we will hold one another as tight as we possibly can.

That healing was so fresh and strong in our arms and hands that we hugged it back into each other, my mother and I, the last moment we would share it, the culmination of the lives we'd lived together, all that pain transformed into something else, something raw and honest and sad and hopeful—not for a cure, not for saving the world, but for a good life in the face of death. We held each other on the airport sidewalk and didn't let go.

"Miss, I hate to ask."

"I know," I whispered into my mother's neck. "I know, soon. Soon." And I cried again.

"I love you so much." She gripped me tighter. "I'm so sorry."

"Me too. I'm so sorry. I'm so sorry. I love you."

"Miss?"

"I know. I know." I let go just a little. I knew this was the last breath before plunging into the dark water we faced, the last gulp of air. The peace would never be permanent. I knew then it didn't matter. It could come like waves, peace and joy, mourning and healing, "I love you, Mom."

She let go a little too. "I'll call you when I land. Peggy is picking me up."

"So, you have a safe way home?"

"I do. I have a safe way home."

"OK." Behind her, a skycap came up with a wheelchair. She let go of the embrace, and I took her hand as she sat down.

The sliding doors opened. "I love you, Monkey."

"I love you, Mom."

Since the terrifying year of 2017, with 20 percent of the population of right whales sick, injured, or killed, there have also been births. While there are only seventy breeding females estimated to remain on planet Earth, the 2022–2023 calving season did see the birth of twelve calves, a baby boom in the right whale community compared to 2017 when no known calves were born. Scientists speculate that whales are feeding more as they change their culture to follow their food to new places. But the future of right whales depends on the lives of krill. And the lives of all of them depend on us. They whisper to their children through the danger as I whisper to mine.

Hamilton told me in our call that the whole problem may not come down to fishing gear or restrictions in the end—not if governments that border the North Atlantic range of the right whale were willing to invest in the future lives of fishermen. With modernized fleets and buoys that use geolocation to track zooplankton, lobster, and whales as well as the movements of the fishing fleet, real time changes could be made to marine traffic, catches could be monitored along with the emergence of the whales' prey, and both entanglements and strikes could be prevented. But that change would require an emphasis on both nonhuman kin and working-class people on the part of legislators. It would require a focus on the vulnerable and a will to support them.

Philip Hamilton began his career in an era of observation, not advocacy. "I wish that with that was coming more protections [for the whales], but I don't think that's the case. Personally, after I'd been in the field for ten years, I'd done a lot of fieldwork. I was at the peak of holding data in my head. We came upon a group of whales; every whale that pops up, I know it. I remember feeling— *we've got this*, we have it dialed in. Any time in science that you have this, the more you look, the more you know you don't know. Oh, I don't actually know that! As you see patterns, some of the patterns of nature take ten to twenty to thirty years, seeing changes—*oh, this is a lot more complicated*."

The work changed him, the proximity to death. "Now I sit," he told me, "very happily, holding the mystery. I feel a lot humbler— a lot more humility—and it makes me more curious. It is openhearted and open-minded."

Maybe that is the hardest work we do, staying openhearted in the face of loss, staying open-minded.

On Monhegan, my mother and Dex and I walked to The Novelty when we were done with our pizza. Dex steadied me on the way, letting me lean on his shoulder. My mother walked slow. Above us the sun began to dip toward the mainland, lighting up Muscongus Bay till it shimmered on the other side of the town, the harbor, Manana. We sat at a picnic table with ice creams. We walked to the beach, where the town and the tourists had gathered in the twilight. We were like any family, mother, daughter, grandson. Seagulls ate the bits from lobster shells dumped on the jetty. We could see the dock and the fishermen on the dock and the sailboats going out from their moorings, and the boats going out of the harbor to watch the sunset.

"You should rest," my mother said. "Sit down."

I sat on the damp sand, still warm from the day's sun. My mother took Dex's hand. The two of them walked to the jetty. I could see her showing him where the mussels and barnacles had found habitats in the pile of rocks. They walked to the edge of the waves. There, in the liminal space between ocean and shore, they picked up periwinkle shells, bits of sea glass, bits of sea pottery. They walked into the water in their bare feet. Dex bent at the waist, looking for animals in the shallows. I could only hear bits of what they said to each other. There was a peace in the way they moved together, a peace I would always want to share with my mother, that I had been searching for my whole life. I taught myself to be the mother whispering to my child in the dark, dangerous seas of our lives. I wished my mother had whispered to me when I was small. Her fear had made that impossible. My fear gave my whispers urgency. Maybe Dex would whisper to his own children in a way that truly gave them peace.

I closed my eyes and let the last rays of the sunset warm my face. I opened them and looked past the harbor entrance, past Manana, out to Muscongus Bay. Somewhere out there were seals and porpoises and lobsters and squid and copepods and all the other creatures, settled after the roiling up of the nor'easter. Somewhere out there were right whales.

American Horseshoe Crab

(*Limulus polyphemus*)

In 2019, Andy and Dex and I went to Monhegan, and then drove south before we drove home, south to the Sound, south to Dogshit Beach, south with my mother's ashes in the trunk.

"I'm not sure where it is." The light was starting to change, a rainstorm coming in from New York City that we were tracking on our weather apps, the roads around Dogshit Beach narrow and crowded. In 1980, the parades of snail shell cottages were low, stretching from unpaved streets to the sea. By 2019, the roads were paved, the plots ballooned, cancerous luxury homes replacing tiny ones, enormous and out of context. They looked vulnerable to everything—sea level rise, hurricanes, financial collapse.

"There's a beach park," Andy said, pointing to a building with a parking lot, *Fairfield Town Beach* on the sign. "Maybe they'll know where to go." Outside there was a booth with a young woman sitting on a folding chair. She wore shorts, a sweatshirt with a hood, her blond hair in a ponytail. I pulled into the entrance and rolled down my window.

"Town parking is $25," she said, a patrician teenage drawl full of vocal fry and uptalk, as if the statement was a question.

"No," I said, "We're looking for Penfield Reef. It's near here, and I think there's public access to it, like a path or something."

"A reef?" she drawled.

"Yeah, Penfield Reef. A long flat reef—stones, shells . . ." I trailed off.

"I think I would know about that," she said, the word *know* a certainty, the word *that* a rising tide of derision.

"OK. We don't need to park if it isn't here." I started backing out. "Thanks anyway."

She shouted in our window, "You mean the rocks over there?" She pointed to the water back the way we came. "That long thing of rock?"

"Yeah. That's it."

"That's a *reef?*"

"Yup," I said, "that's a reef," and I rolled up the window.

"Weird," said Dex from the back seat.

"Let's drive back the way we came."

We turned around on the narrow road, drove back to a corner with a parking meter, a convenience store, a bar opposite. We parked. There was a crosswalk that looked like it led between the beach houses.

We took my mother's ashes out of the trunk where they had been the whole trip to Maine, the whole drive back to her old house in the woods in Massachusetts where we had the memorial with her friends and her sister. We took out the heavy box, the unmistakable box—the one we carried onto Monhegan in our luggage so she could be with us there one last time—and put it into a tote bag. I carried it over my shoulder. That last walk she was my burden. We went toward the break in the houses.

"I used to go somewhere over here for ice cream when I was little. There was a general store, and the beach road was all sand. There was a mean dog chained in one of these yards."

We walked on. The clouds were ominous, weighty on the horizon, but overhead it was still blue. We found a public access point. *Penfield Reef,* the sign said, with a list of rules for the place, none of which we read. There was a boardwalk, then a long, flat, stone jetty I remembered from childhood, covered over with barnacles and mussels, periwinkles between the stones.

In 1980, I fit my fingers into the spaces between the rocks of that jetty to find the purple shells of live mussels and the sharp surfaces of a hundred white barnacle mouths. I pulled at a single mussel, bracing my foot against her rock, and tugged until she came free. Deprived of seawater and removed from her colony, the mussel seemed to lose what little animation she had to begin with. She withered, she shrank, and I threw her out into the surf for the gulls, ashamed I'd killed her magic. But a marine mussel's byssal thread can regenerate, returning the broken to wholeness. Many mussel species have byssal threads. A byssal thread is a footlike appendage that is flexible, extending from the animal down to become its anchor on any surface it chooses for a home. The Atlantic mussel uses the byssus for locomotion and permanent anchoring. The byssus that extends from the mussel's shell is a collection of threads—made of stranded proteins and studied both for their adhesive properties and for their flexibility—and really more like a rope, grown from a byssal gland inside the creature's foot. When the byssal thread is broken—by storm or predator, by a heedless child, by chance or intention—they can rebuild the foot that connects them to the rock, using a cooperative healing process to make themselves whole.

At the end of the jetty, the reef began, all shells and loose rocks, pointing out into the Sound like a labyrinth that had been uncurled, the far end a thin tail. We began to walk it as the rain picked up. I could feel the tension in the three of us, resentment at all this work to scatter her here, a place we'd likely never come to again, a place we didn't visit. We might have sent her ashes from the quiet of Lobster Cove, or into a waterfall on a coastal Monhegan trail, or over a cliff she was never brave enough to stand atop. But she'd asked us to go back to Penfield.

"She was a pain in the ass when she was alive," I said, half smiling at Andy and Dex.

"And she's a pain in the ass now," Dex finished.

"Had to make it rain, huh, Nancy?" Andy asked.

I hugged the bag closer to my body, my hair blowing. I tried to take pictures to remember, to send to her sister, her friends. It was only the three of us.

Out on the reef, a man stood fishing with his feet in the water on the windward side. I worried he'd see us—it was illegal to dump ashes. It was like my mother was speaking the words in my ears. Her fear pushed in at everything. *What if we get it wrong? What if there was a perfect way to meet the moment?* Then the wind picked up and the bag bounced against my flank and I felt the pain in my kidney, and I was jolted out of it. *Everything is already perfect and wrong, ruined and amazing,* I thought.

We kept walking. Halfway out the reef we stopped. The view to the windward side of the utterly flat reef was prettier. It looked like open ocean there, facing the Narrows; the beach was clean. If we turned and stood with the wind at our backs, the Sound looked like a sulking pond, barely a wave, deadened, the Bridgeport factories in the distance. But there was no way to empty the bag on the pretty side, the wind blowing strong in advance of the storm. We opened the bag, wind at our backs, facing the homely water where I'd tried to learn to swim, each of us taking a turn to scatter her. I tried going to the pretty side, but the wind came up and we got ashes on our shoes. We paced around, putting shells in our pockets, until the bag was empty, and the rain arrived.

"Should we run back to the car?" Dex asked.

"What did we learn at the science museum about how raindrops work?"

"You don't stay any drier if you run."

"Yeah, let's take our time," Andy told him.

The fisherman never even looked up. "Let's remember one thing that she did that was a pain in the ass," I said, walking fast, head down, "and one thing that was great."

"She was afraid of everything," Dex said. "And she was always worried."

"And a good thing?" I asked.

"She loved me."

"She was very anxious sometimes," Andy said. "But she welcomed me, that was great."

I said, "She never thought I forgave her." I looked back along the sandbars and jetties, to the huge new houses with their upper decks and crowded plots, where The Periwinkle had been. "But she taught me how to love science and creatures, and she let me be curious."

"'Me too," said Dex.

"And she worked to save us all."

We all looked back at the reef, the rain plodding—big drops, cold and intermittent—and we stopped hurrying. On the edge of the waterline, there was movement.

"Mom, did you see that?"

"Yeah."

Dex walked to where the movement had been and bent down. He stooped and turned to us holding an enormous horseshoe crab, old, her feet wriggling. "Look!"

He stood there, my child, fourteen and sure he'd study sea creatures for the rest of his life. He held up the horseshoe crab, his feet on the reef where I spent my childhood. He was alive with it, even though the ashes were still on all our shoes, the rain on our faces.

"Take my picture! With the crab! I can't believe it; I've always wanted to see one."

He stood there, my child, and held the crab by the sharp sword of a tail. Horseshoe crabs have been on Earth for more than 480 million years, longer than dinosaurs. Horseshoe crabs are essential for medical science. Captured and taken to labs, their blue blood is extracted, then used for its complex ability to clot dramatically in the presence of bacterial toxins. In vaccine development, the blood can tell scientists whether the vaccine is safe or not. We bleed them, then return them to the sea. We rely on them. They are also under pressure in a warming world. Without them we have no vaccine science. Without them shorebirds have no horseshoe crab eggs to eat. Synthetic versions of their blood are in development. Their future is bound to us, and to all living things, and we are bound to them.

Dex looked back at the camera, afraid of nothing—not the storm coming at us from the west, not the whirling legs of the prehistoric monster in his hands, not my health, not his hopes, not the tide rising on the reef, not the ashes on his shoes.

On that trip to the Cape with my mother that summer he was six, Dex found a pool of the molted carapaces of baby horseshoe crabs, bleached in the sun. We took home fifty of them, and he lined them up on the railing of our cottage, a parade echoing their ancestors and new science about the trilobite, that they often migrated in a similar parade, head-to-tail, looking for food, then came to a place they deemed safe, and molted their carapaces and looked for mates.

Layers and layers of time and truth and history and future and feeling were flashing all around him as he held that horseshoe crab up and watched her move. It was dizzying, his courage. He had no need to look away—only into the camera—ready for the world to come.

When she was dying in the hospice, my mother's meditation teacher came to sit at her bedside. He was a Tibetan monk, and she studied with him for years, trying to undo what she had done, to my father, to me, to herself, trying to learn to live with the whirlpool. She found something I couldn't see then: peace. As she left, she was losing that peace, so he came and he held her hands and touched her forehead with his and reminded her that this transition was just a new phase. "You'll have a new body! A brand-new body!" he said.

Then he told her that in the next life, all the things she'd done and learned and studied would be in that new body like seeds. "I had the seed for Buddhism," he joked, "but not the seed for English! In your next life, when you hear *Buddhism, geology, environment, water*, then the seed, it will wake up! Your conscious mind will begin to remember, and you'll go faster toward what you want to know, with the things you already learned in the seed, growing!"

My mother's face softened when he said that, evened out, and she looked calm for the first time in days. All she understood, and loved, and worked for—water protections, meditation, compassion mind, stewardship—all of it would wake up and come alive again, in that new way, in that new body.

On Monhegan—the year of the nor'easter and the seizure and the hospital, the year of my hypoxia, the year I had to say out loud that I was going to fail if things didn't change—the luggage truck pulled up in front of Holdfast House on our last morning.

"Truck's here!" Dex called from the porch. We went to meet it, our bags and recycling and coolers carefully arranged to be easy to load out. There was fog over the island.

Standing in front of Holdfast House as the truck backed up, my mother said, "I never got to do this for you. I never found a way to get to Monhegan when you wanted to come. I never got to share it."

The truck parked at the base of the walkway, and the driver stepped out—the young firefighter who had brought us emergency clothes and stuffed animals in the rain the night of the accident. "You made it back!" he said, interrupting.

My mother hadn't been able to do what Mac had done. She had done something else, something she didn't notice she was doing, or remember. She was trying to get it right, to repeat a time she remembered as happy. She missed what we could have had, even with my father sick, even with me sick, even with us poor, even in 1980. While she was dying in 2018, the presidential administration began taking apart the Clean Water Act, the Clean Air Act, the Endangered Species Act. When my mother was first diagnosed with cancer, she told me she wanted to live to see the pendulum swing back. "My entire life's work is being dismantled," she said. Years after her death, a different administration created legislation that could fundamentally alter the way we respond to ecocollapse, and the lives of working-class people caught in its jaws and the lives of nonhuman kin swept up in its tides. The pendulum keeps swinging.

The ocean has no beginning, no middle, no end. It is a spiral of the water cycle and the currents of the world. The body has no beginning and no end. It is a spiral loop through ancestors and DNA, through exposure and spirit, epigenetics and the future in the cells of the future people you might create, even if you don't create them. That futurity, that history, hangs on us as beings, endless and cyclical, like the endless round of birth and death my mother meditated to free herself from every day that she sat at her altar. She sat on the prayer cushion and stirred the singing bowl and held the beads and tried to quiet the pain in her joints and bones, the pain of fibromyalgia and the cancer that was brewing somewhere, of the depression, of the chronic fatigue, and prayed to be one step closer to release from the suffering of the world. She prayed for the release of all beings from suffering. The sea is endless, and so is the body, and so is mourning, and so is the endless round of suffering common to all creatures; it is the story of the waves, the water cycle, the beings caught in both, myself included.

I am not free of mistakes or trauma or repeated patterns of working to become free. I will not be free in this life, even if I pray every day from now on, even if I parent as well as I possibly can, and tend my marriage as well as I possibly can, and become as open as I possibly can. I will move only inches. But within those inches is everything I will need to begin, in this lifetime I am offered, to come closer to healing. It is more than a story where I am returned whole and perfect from a life of struggle. That is not what happens to people, returning perfect. No recovery, no healing, ever led to perfect health, and to wish for that is a way to spin the wheel of loss and return and suffering for another round, if not in your own life, then in the lives of those who follow you.

In early bestiaries, flying fish were drawn to enormous scale in illustrations, with pantherlike heads and wings that spanned the size of the boats they menaced. The terror of mariners, their sharp sails were reported to cut people and boats. Between 1215 and 1220 CE, Vincent of Beauvais, a Dominican friar, wrote his bestiary-cum-natural history encyclopedia *Speculum Natuerae*,

"Nature's Mirror." In that book, he calls flying fish sawfish for their sharp spiny fins and wings. He wrote, "The sawfish swimming hidden beneath the ship cuts through its bottom, so that as the water rushes in, it drowns the crew by its crafty device and gorges itself on their flesh."

In a later volume—the thirteenth-century bestiary simply referred to as the *Sloane Manuscript*—this entry is explained: "Now the sea is a symbol of this world. The ship is a type of righteous persons, who without peril or shipwreck of their faith pass through the midst of the storms and tempests of this world (and overcome the deadly waves, that is, the adverse forces of this world). But the saw-fish, that is that beast which availed not to beat the ship in sailing, affords a symbol of those persons who at first eagerly engage in good works, but who afterwards do not persevere in them, and are led astray by faults of different kinds (that is, of greed, pride, drunkenness, and luxury), which toss them about as it were upon the waves of the sea and plunge them down to the depths of hell. For not to those who only make a beginning, but to those who persevere, is the reward promised."

We are endlessly thrown down into the waves and maybe into hell as we attempt to rise above what has harmed and hampered us. We have our faults to tempt us back to the waves. Perseverance, or perhaps practice, provides the only way forward for the ship, the righteous person caught in the sea of this world, me holding steady over the vortex.

In 1987, the year the Long Island Sound turned hypoxic, anoxic, and nearly had the majority of its sea life extirpated, a flying fish, caught in the Gulf Stream and pushed far north of its usual habitat, was landed not far from its waters. Other fish have been caught in the Atlantic off Long Island, off New York's harbor that do not belong where they have been found. In a 1988 *New*

York Times article, John Waldman, then a research assistant with the Hudson River Foundation, listed species he'd collected himself and those that had been reported from Montauk and Gardiners Bay, Orient Point and Jones Beach, Shinnecock Bay and Sheepshead Bay and even in the Hudson. Warmer-water fishes are found there—sailfish and trigger fish, amberjack, snook, squirrelfish and foureye butterflyfish, manta rays, pompano, lookdowns, tarpon, the great barracuda. "All signs suggest this is a particularly good year to search for tropical fishes. This summer's record-breaking heat warmed Long Island's waters to around 80 degrees, and parts of the Hudson River to even higher temperatures," Waldman wrote. The future that was coming had harbingers in those fish, those waves of marine strays.

Since the future is arriving, has been arriving, in waves of marine strays and waves of heat and waves of loss and waves of rebounding, it may seem as if the nature of it tends only toward chaos, death, making mourning a discipline that is forced on us, or a consequential skill we must perfect or be lost ourselves.

But when my mother sat on the prayer cushion asking for her own liberation, she was not assuming she would find it in this lifetime, or even believing she knew exactly what it would resemble. She was making a channel in herself, slowly wearing away the things that blocked her in hopes of an opening, a moment of grace that gave her a vision of the world she wanted, the love she wanted, the future she wanted, not only for herself but for me, for Dex. She was looking to abolish her pain and failure and resistance, the way she'd leapt toward grace and goodness and god, falling back into the water, a flying fish, a leaping sawfish, somewhere between liberation and the work it takes to get there.

It is only possible to imagine another kind of future when we accept mourning and recognize that the future we are offered when vulnerable—the exit of surrender to death or to relegation as a lesser kind of human, one not worth saving—is offered out of fear for what transformation would make in a new world. We go through waves when conventional wisdom treats resources individually—water, air, soil, minerals, fossil fuels. But they are a web of stories, interactions, inadvertent consequences, and chance results. We have to see the whole of the problem at a diagnostic level and think holistically about what constitutes a real healing, deeper than regulations that can be overturned, deeper than a single species that we notice in the daylight, deep down at the root of the problem. Spiderweb thinking emerges in some human beings—the naturalist, those with unusual minds, those outside of the norm through illness or difference or loss. It is a way of seeing those connections, and understanding the possible, wild, joyful, dancing paths to heal them. Mourning is a template for that, a spiral like the nautilus shells glued to Mac's sliding doors, chambered histories revisited as a life, a loss, a way forward that twists around itself, returned to again and again.

At Holdfast House, our luggage was loaded into the truck, and the truck took it to the harbor, and the boat hands put it on the ferry. This ferry was the *Laura B*, the deliverer of Dex and me that first time we'd seen Monhegan in 2008, the supplier of a working island, with bundled packages on the deck and seats close to the water.

On the other side of that water—somewhere between the harbor on the mainland, somewhere between my mother grinding the gears of my stick shift all the way down Route 1 and the moment we were delivered by Martha back to our apartment in Chicago, somewhere in the middle of dizziness and sleepless nights and Dex coming down with pneumonia from the stress and the hospital—something changed.

I healed from the accident, as I had started to do standing in the kitchen of Holdfast House with my mother, as I had watched her hold Dex's hand on Fish Beach. I would never have another seizure. I would wear a heart monitor and have EKGs and EEGs, only to find out that my growing kidneys meant that when I tripped on a single small rock on a dirt road on an island in Maine twelve miles out to sea in a nor'easter, I fell onto a body already struggling with kidneys that were crowding out my other organs. Those kidneys made it harder for my lungs to expand, making singing difficult, making it so that I had less air in my body, making it so that the hard fall pushed all the tiny volume of air out of me, and I became briefly anoxic, just like the Sound. Starved of air, I seized.

I would never be able to make myself into the person who could soothe my mother or keep her from guilt over the things she couldn't save. I would never be able to be well, not as she and I both wished I could be, delivered from the genetics of the Caffall Curse and the history of my father and all my mother wanted to put behind her. I would never be the pristine sea teeming with fish and the clear river rolling into it, full of trout and salmon and nutrients and cold water. I would be something else. I would be an ocean that could be made whole, not immediately, but over time, with care and volunteers, with boring regulation and break-through science, with local leadership and Indigenous collaboration, with understanding the full story and telling the whole truth and with care and love especially.

As Martha drove us home, the three of us would sing "Southern Cross" at the top of our voices, and the yacht rock would swirl around us. *Think about how many times I have fallen,* we sang. And I would remember that I had music and people and a place to go home to. I would call Amanda and Lawrence and Robin and

Laurie Lee and Shana and Molly and Karen and every friend I could reach and tell them, "I'm in trouble." Somewhere in the Midwest, they would begin to work on my behalf, and within the year they would throw a fundraiser to pay for the cost of the accident. "Parachute Among Anchors" they would call it, after a song I wrote.

The fundraiser was the parachute that helped me escape; the fundraiser was the parachute that brought me softly back to Earth. My accident was the anchor that held me to sorrow; my accident was the anchor that held me to being human. I would begin to work on my own, and slowly—sharks coming back for seals coming back for fish—things would begin to change.

Jonah wouldn't know how bad it had all been. From the road, once I was safe with Martha, I called him and told him I'd been sick, and he talked with Dex about it. He would never help with the illness, but I would never be gripped by fear of him like I had been. He was not going to take Dex from me, not going to steal the time we needed. There was time, even time for Dex to understand how to be his child. I recovered from my seizure, trying to form a life for Dex and me, living near friends, slowing down. I began to play again. Lawrence and I worked together in the band. We learned how to be friends. We learned how to talk about what failed. We let it go. Dex learned to lean on him as family. I began to date again. I rebuilt. It was slow. I fell in love with Andy, and I married him, and we lived together with Dex in a house full of fossils and cats and books. Dex found in him another father to add to the aunts and uncles who surrounded us in the years after the accident. Dex learned to have a relationship with his father; he is learning still. Jonah married again and had another child. I tried to heal all that had been broken. I am trying still. My mother sickened and died and, finally, in the water of the Sound, the last of my mother's body joined the current that had been joined by her mother and father, that had been joined by my father as his ashes came down from the Konkapot and the Housatonic. There, in the wild swirling of the planet's oceans, they are still, combined and reborn, joining the largest engine on Earth, the life and blood of us all, forever and ever, amen.

But, before all of that came, when the hope of slow repair was just an idea, my mother stood with Dex at the bow of the *Laura B* as it left the Monhegan harbor. They stood on each side of the flagpole there, holding hands, hunting the waves for animals. In the distance—along the smooth surface of Muscongus Bay—we could see the unmistakable curved lines of the backs of a right whale mother and calf. Dex took my hand, and my mother took my hand, and we watched them. Something started changing in me, something that I had forgotten. If the cycles of repeating mourning were spread out around me that day, along with them, intertwined confusingly and inextricably, were cycles of healing. Like waves, like concentric rings, like ripples from a stone thrown into a still sea.

I looked at the whales and whispered to Dex, "Isn't it powerful?"

MONHEGAN ISLAND, 2021

Every text remains in mourning
until it is translated.

—JACQUES DERRIDA

Humpback Whale

(*Megaptera novaeangliae*)

On Monhegan, Dex and I walked to Swim Beach one evening, seven years after the night of the nor'easter. We stood on the edge of the waves. Clouds bellowed out on the wind, shading the sea below them, great open sails of white and gray, hazy rain shadows connecting the water in them to the water they fed—a cycle, a spiral, a link.

"Should we swim?" Dex asked, face lit by sunset, shaped like my father's and mine.

"Storm's coming."

"It's not here yet." He turned his head and smiled at me, one corner up like a dare.

"OK," I said, "maybe we have time."

"Yeah," he said, "we've got time."

We kicked off pants and T-shirts, down to bathing suits we'd hidden underneath. We ran at the cold sea like my mother had at the Sound, like Mac and June and me skinny-dipping in the bioluminescence, like Dex and me in Lake Michigan, like a thousand moments when I'd run to the water for comfort or truth or god. The cold of it caught my breath hard as I submerged, then popped up and ran back to the shore, "Cold! Cold! Cold!" I screamed, until I was back on the beach.

Dex bobbed in the water looking at me. "Come back, Mom!"

The waves kicked up and Dex let out a whoop, throwing himself into the foam, body surfing toward the shore. I laughed, walking back into the sea, my feet just touching the sand and stones, my toes gripping, watching him, his joy, his fearlessness. I said a mantra, *The water isn't that deep, the water isn't that cold, the water will heal your body.* I was tired, but water is what I love best, and what I know better than anything else, outside my skin and inside my skin, moving in a moving element.

I dove in to join him. My mother was with me in the water, and Mac, and my father, and Aunt Lois. In the Sound, decades ago, teaching me to swim, Mac said, "Find that balance in your muscles. Remember if you feel like you can't breathe, you can take your time. Everything should work in tandem, in concert, your lungs and your arms and your legs. You should feel like you are at one with the water. It is you."

All my ancestors were with me, two centuries of drowning people, and all the doctors who have ever helped me find healing. There was the acupuncturist who told me in 1999, "I can keep you away from dialysis for a long time. The Chinese are obsessed with the kidneys."

There was the nephrologist who said, "Well, I don't know anything about acupuncture, but if it works, do that!"

There was Dr. A. "You may never lose kidney function if you can just hold on. There are new treatments every year." There was Dr. A teaching young residents how PKD changes lives, bringing in patients, myself included, to tell them about the cost of it. "Remember," Dr. A said, "it isn't only the patient who walks into the exam room. They come with all those family members too."

I was born into a family that has an illness that has followed us for at least 150 years. But every day there is more hope in the treatment of that illness.

Zebra fish don't live in the Gulf or in the Sound. They live in the shallow, fresh waters of Southern Asia, breeding. But in recent years, zebra fish have acquired celebrity status in the field of medical research. They have what is known as *high genetic conservation with humans*, meaning that nearly 71 percent of their genes are shared with human beings. Because of this fact, as well as the remarkable adaptations that they have of their own accord—they are transparent and can heal and regenerate in remarkable ways— zebra fish have been the subject of numerous medical studies to understand disease progression and cure. In the study of kidney disease, zebra fish sweeten the pot of scientific discovery with a strange trick—they can regenerate nephrons, those internal mechanisms that aid the kidney in transferring fluid between the circulatory system and the grapelike glomeruli that act as the body's filters inside the structure of the organs. Zebra fish can repair their nephrons. Zebra fish can generate new nephrons throughout their lifetimes. Zebra fish might hold the key to the repair of kidneys damaged through PKD, or at the very least, the key to understanding how to treat them medically.

Zebra fish are in the grips of something as well. Even as we study and manipulate them, even as we hope they will provide answers to human mysteries, human actions are destroying them, and the answers besides. The warming waters of Southeast Asia are proving increasingly inhospitable to zebra fish. In a December 2020 article published in the *Proceedings of the National Academy of the Sciences of the United States of America* (PNAS), Rachael Morgan and colleagues presented findings from a study of zebra fish. They selected the offspring of wild-caught zebra fish and in one group of them, subjected six generations to increased water temperatures. They did this over the generations, raising the temperatures slowly to see if the fish could acclimate over time, but they couldn't, rates of acclimation were minuscule. "These results suggest," they wrote, "the existence of a hard limit in upper thermal tolerance. Considering the rate at which global temperatures are increasing, the observed rates of adaptation and the possible hard limit in upper thermal tolerance suggest a low potential for evolutionary rescue in fish living at the edge of their thermal limits." In other words, zebra fish cannot adapt fast enough to meet the rapid rise in global water temperatures. In six generations, the zebra fish could not adapt to the increasing heat of climate collapse. Like most vertebrates, their ability to change was slow. Like most families, ours was too.

In the march of climate collapse, while heat and extreme weather patterns are more likely as the planet warms, heat waves, including marine heat waves, have their likelihoods increased exponentially. Meaning that the Gulf, the Sound, and the South Asian homes of zebra fish are more likely to suffer loss, collapse, the extirpation of species, the end of generations far beyond those six studied in the lab.

According to a study from World Weather Attribution, reported in part by Scripps Institution of Oceanography in July 2021—just after the devastating heat wave that exterminated billions of mollusks on the Oregon Coast, that boiled salmon in their rivers—those deaths, that heat wave, would have been impossible without climate collapse. In the study, "An international team of 27 scientists calculated that climate change increased chances of the extreme heat occurring by at least 150 times, but likely much more," according to reporting by the Associated Press.

The treatment of PKD has seen a steady upward climb of technology, but if the world were to collapse, all of it would be lost without the electricity that powers dialysis, the manufacture of medicine, and medical equipment. The ill are one hurricane, one supply chain failure away from death. Monhegan protects itself, sustaining its private land trust of nearly four hundred acres, currently managed by the Monhegan Associates Incorporated, creating The Monhegan Island Sustainable Community Association (MISCA), an organization that raises funds to sustain affordable housing for year-round residents, providing lower rental costs, and helping residents buy land and property to support businesses. The pressures of the outside world encroach on the island just the same. The water is warming everywhere in Muscongus Bay, and in the last decade, the company New England Aqua Ventus wanted to erect one of the world's largest offshore wind turbines right in the heart of the lobstering grounds of Monhegan, a move made possible precisely by the fact that the relatively small number of fishing families on the island don't have the political clout, or the numbers, to push back that mainland lobstermen do. A working pickup truck on the island flies a flag from its bed—*SAY NO TO OFFSHORE WIND, PROTECT OUR OCEANS*. Mainland lobstermen joined the Monhegan families in a more-than-eighty-boat blockade of the project. The project appears to have moved on to other waters, but the pressure revealed again the deep vulnerability of the island.

There is trapped heat in the Gulf and in the Sound and in the homes of zebra fish. And we must hurry to fix it, if it can be fixed at all. We must recognize the wickedness of wasted time, the urgency and fear. But, like my friend was told when she was diagnosed, we must remember to dance. Like me when I was diagnosed, we must remember to sing. Like Dex and me and the periwinkles, we must remember it is a miracle that we exist at all.

Dex was with me in the hospital after the nor'easter; he saw how things changed after that night, the new doctor, the new family, all the ways in which I was trying to make things better for him. He knew the stakes. The pressure of my father's fear and anxiety had driven me to get tested young. I know that the changes I made then have extended my physical health. But knowing your future is something you should choose only when you are ready for the emotional questions that come after, the spiritual work that follows, the conversations with god. Dex will have to know when the time is right for him.

Dr. A begins each appointment asking me whether the reflux and lowered lung capacity from my growing kidneys has interfered with the things I love.

"Are you still singing?"

"Yes, I'm still singing."

In a recent visit, I asked Dr. A about the RNA treatments that were being tested on mice, a potential cure for PKD. "Oh," she said, "we had to stop those trials. The drug melted the cysts from the bodies of mice, but not from humans. It seemed really promising."

"Oh, that's hard news," I said.

"I know," she said, "but we'll find something. There are a lot of people working on it." She paused, "Do you want to get Dex tested? I know this is complicated for PKD families, but the sooner we know about Dex, the more treatments we can offer. It isn't at all like when you were his age."

"My diagnosis was really upsetting."

"We just have so much more we can do."

In a February 2019 podcast for physicians about PKD from the University of North Carolina, Dr. Amy Mottl spoke with Dr. Robert Falk about the nature of PKD and the process of diagnosis. In it, Dr. Falk said that diagnosis in one's teens and twenties was contraindicated. Knowing early would only lead to denial of health insurance, and since there was no known treatment or cure, the resulting emotional trauma and anxiety, especially for patients who had witnessed the illness or death of a parent with PKD, was too high a risk for the small reward of early detection. I am not sure if my early diagnosis helped. But watching the seamless switching from wonder to extinction Dex has always practiced, that I have tried to practice in the years since my diagnosis, I feel there may be some benefit.

"Dr. A keeps asking when you want to find out if you have PKD," I told him once.

"Not yet," he said. "Not for a while."

"OK," I said. "I'll tell her."

I taught him to think of PKD in a way my parents hadn't been able to teach me. He learned that families can be hard, and how to survive his father's rages. He learned families can be soft and full of joy. He learned to swim in lessons at the public pool, where I would swim after he was done, and he would sit on the pool bleachers with his homework. A string of thought connected me to Dex, learning biology so that he could study fish one day, or invertebrate marine animals—filter feeders, the kidneys of the sea.

In the pool, I thought about helpful ghosts, and I kicked harder, in a style cobbled together from Mac's lessons and afternoons at the lakeshore chasing Dex and patient days of practice. I moved faster through the water. I heard Mac's voice, felt the warm ripple of my wake, just like in the Sound, all around me.

At the edge of the pool, I stopped swimming to shake water out of my goggles. I adjusted the kickboard I used to steady my trunk and protect my kidneys from twisting as I swam; the familiar constant of flank pain from my kidneys came back if I didn't. The woman in the next lane said, "You're a kicker!" She pointed to my kickboard. "You're lapping the rest of us using just your legs. It's annoying, actually."

"Oh, I don't know what I'm doing!" I said. But I realized that I did. I was finding what worked, remembering the dead. Moving in a moving element.

In the Gulf, that evening as Dex and I walked into the cold Monhegan harbor, there were animals holding on in the warming water, just like me in my flooding body, just like the animals in the once hypoxic Sound. If the worst comes, it doesn't always mean the end; you may spend your life trying to fight against it or succumb to sorrow, but the grief could send you on a path toward mourning that is joyful as well as somber, full of wonder and the ability to hold the sorrowful future and its redemption simultaneously as you move forward.

The power of a bestiary is the same as the power of illness and the power of collapse. The whale can be an animal alone, she can be the charismatic megafauna that stands in for the loss of habitat, she can be an island that pulls us into the depths. Loss can be the cursed body, it can be the opportunity for cure, it can be the door to transformation. We are washed together in the same ocean—blood to water, dinoflagellates to cells, toxicants to toxins, over-enrichment to kidney failure, ecosystems to bodies—in systems that can be healed only by insisting that every life caught in them is worth saving. Telling their stories, telling our own, opens a door.

At the end of the children's book *Grover and the Everything in the Whole Wide World Museum*, Grover the Muppet spends all day in a museum—in rooms of soft things, hard things, things in the ground, things underwater, heavy things, light things, and cute things, where he has his own pedestal. The last room has a huge, arched door. Above the door are the words *Everything Else*. When he opens it, there is the world.

If there was a door at the end of my bestiary, above it would be the words *All the Other Creatures*. This book lists only a few animals, covers only a few ecosystems, all of them aquatic, creatures that feel like kin to me, because I have seen them and known them over a lifetime. Bestiaries are not meant to be exhaustive. They are meant to be enlightening. We all make our own. We see the insects in the garden, the cranes in the marsh, the lizards on our porch. We have our own secret names for them. We have our own stories of what they mean. I swam in the Sound every year for a decade. I swam in the Gulf every year for a decade. I still know nothing about these oceans. They are full of mystery and strength, discovery and promise. I will not live long enough to do anything but hold them in awe. I will rest in the unknowing. There is hard work. There are failures and returns. There are new threats and bad laws and attempts at justice that fail and attempts at justice that stick. There are waves and waves. The whirlpool is always waiting, and so is the spray of stars on the other side.

In the Sound, humpback whales returned in 2015. The water has been recovering for thirty years from the hypoxia that almost extirpated all its life in 1987, the year we scattered Mac in the waves. The menhaden came back. The alewives came back. People foraged for littleneck clams. A tiny fish I never knew as a child, the sand lance, came back.

Sand lances are thin and silver; they burrow in the sand of the flats and the floor of the sea to escape predation, deep down where I used to dig my toes to keep from fearing the depths. They are insufficiently studied and under-regarded, hiding there where the sand is safe and cool, away from the vast open sea. But the humpbacks love them and come back for them. They love them as they love returning menhaden and herring. The clean water drew the humpbacks home, those whales that are kin to the ones Mac saw from her catamaran in the 1950s, her daughters watching from the shore, feet in the Sound, waiting for the water goddess to come home.

Like the sand lance, drawing in predators to rebalance the sea, there are mysteries waiting to be understood. The rest of the world of creatures is waiting with meaning. They are in your own eco-system, in the ecosystem that isn't yours but that you love best in all the world, in ecosystems you will never visit but that flow and bleed into the ones you know. They are yours; they are mine; they are their own.

I know that the whirlpool of my body and my mourning gave me something rare, a chance to see it as a vortex, a chance to see, in that campground in South Dakota, a way forward that chose my own life, even if the spiral toward despair and doubt was always present. I swim, I feel ashamed of it, I pray, I don't talk about it; I follow the whirlpool toward something true.

On a monastery retreat once, I booked a spiritual counseling session. It took me forty-five minutes to say that at twenty-three, grieving, I'd seen god in the Badlands. It took me longer to say I was angry that I had no container for that encounter yet, no path, chosen or inherited for facing the mystery. The counselor said, "I would give anything to see God."

I have a container now. The world is the container, the Earth, even my body. In 2021, only a few months after we swam in the cold harbor, Dr. A put me on a new drug for PKD. Tolvaptan was approved in the United States by the Food and Drug Administration in 2009. It is only available for adult patients who have entered the "rapidly progressing" period of the disease. A vasopressin antagonist, it works by starving the patient's cells of water, reversing the drowning body, slowing the growth of cysts. This causes extreme thirst, frequent urination, lifestyle changes that can be disruptive, but it can give patients years of kidney function back. It can also cut PKD-related pain by 30 percent and restore lost energy to some patients. Dr. A was one of the leaders of the longitudinal study that helped bring Tolvaptan into public use.

When my kidneys reached six times their normal size in 2021, it was time for me to try the drug, the very pill my parents mentioned when they told me that, one day, I might have what my father had. It wasn't a cure, but it was a healing. Within days of starting the drug, ten years of constant pain ebbed, and the energy I remembered from my thirties, when the disease was a distant threat, returned. I think of water all the time now, dying for it, craving it, wondering where it will come from next. I am drowning more slowly now. I am getting used to the idea that I will live. I am wondering what it would look like if we tried to get used to the possibility that the places we love might survive too. I am holding the idea of death and life at the same time. I am, as ever, a water creature. I am still in collapse. PKD stalks me. But I have time, time to wait out the storms coming, to find solutions, to hope, to live.

I have been writing this ocean elegy for a long time. I have felt bereft and lost sleep and read studies of the deep ocean and reports of plastic in bellies of seabirds and wondered how to be of use. I have watched scientists and thinkers pale under new information, and students and children break open when they understand. And I have tried to catch the fog of it in a specimen jar and say, *Now, now, now, I know what this story is, now I know it is about the deep ocean, or the fastest warming ocean, or the estuary.* I have worried that when the fog is just captured and I regard it through the specimen jar, it fades and dissipates, that the animal I wanted to tell you about is no longer here, or no longer threatened, or no longer abundant since the abundance predicted its crash. When this book is finished and published, and then years from that, things will have changed again, and you will say, *Yes, but then there was the change; remember when that thing changed it all?*

Illness is like that too. I cried for joy when my friend with PKD told me about the surgery that took her kidneys away, gave her a new kidney and the immune system of her donor, making her a chimera. In some bestiaries, the chimera was included as a symbol of the raw, satanic power of nature, the ways it could be bent and unnatural. In my friend, this change is a miracle I could never have imagined when I was a teenager watching my father line up antirejection medicine on the kitchen counter, knowing that the American laws at the time only covered these drugs that kept him alive—which even then cost $800 a month— for four years post-transplant, at which time the patient was on his own. There were calls to change the law. The law changed. It may change and change again. I wonder, *How will this story be true tomorrow, or when will illness not be a story at all?* Already it is not the death sentence of my father's time. Holding on was survival.

I want to list all the creatures that are threatened with loss. I want to list all the thousands and thousands of hours of volunteer time and scientific time and activist time that attend every loss, every homely, invisible creature with advocates and specialists sitting at the bedside of their extinction, as I sat at my father's, my uncles', my aunt's, as my child will sit at mine. For every PKD collapse there is a Dr. A—the war stops, the pandemic ends, the Depression eases, the extinction completes or reverses, the dying person slips away, the sick person recovers, the loss begins again. Life is liminal space between health and illness. This story has a finish, but it also has none; surviving is the finish. If things were to end, I could write the elegy. As a permanent mourner, elegy feels premature and required in the same moment.

It may be that we have entered a period where mourning is permanent for all of us. But mourning is not the same as grieving. They are distinct. Grief is the raw element of loss. It is singular, private, individual, internal. Mourning is the external experience of loss, the public face, the collective, the unfinished story we make together. Survivors, whether of genetic illness, or settler colonialism, or class warfare, or systemic racism, or ecosystemic collapse, are not going to stop feeling loss, ever; we are all permanent mourners. I have settled into this truth. I have found it to be the doorway to joy.

Permanent mourning allows us to change and change and change again, evolving a way out to the future. Mourning as I am, I feel tumbling out of my mouth, faster than I can say them, the names of all the animals in these places I love. They are the hundred thousand names of the dead on the cover of a newspaper, etched on a marble war monument, black crepe over the former site of the glacier. We are witnesses, in the dark, doulas for the birth of the world without the dead. Legacy comes with new life in structures you cannot see unless you peel back layers. The overgrowth of algae was a threat to fisheries, to bivalves, to carbon, to swimmers, to copepods in one generation and the next. The overgrowth of cysts were a threat to the pharmacist, his child, to my father—written in epigenetics and memory and sitting on me like a hungry ghost, a mystery, a fear. On the other side there is new life. I have seen the births that follow loss.

The cost of the collapse is hard. The mourning never leaves. But the chance at opening, transcendence, power, change, are all there in the whirlpool if we choose to see them. We belong to one another. We live in a world where nothing is ever separate, nothing is ever over. We swim through the lessons of our bodies in interconnections and vulnerabilities we cannot avoid for as long as there is life. In that beauty is a central truth—there is always transformation, and there is always grief; nothing leaves us, and no one is alone.

Most of us don't know how to mourn that way. But as the world warms and changes, as collapses arrive and are healed, we'll have to make a practice of mourning that changes us, too, one that makes mourning something powerful.

When I was diagnosed, I thought I understood the whirling life I was born into—body surfing when the wave takes you, turns you, skins your knees on the sand, delivers you, upside down, mouth full of brine, grasping toward remedies withheld without reason. It took years to understand that the remedies were always there, waiting with a gentle confidence that they belonged to me already, as long as I was willing to fight to save my life, my precious life.

It takes the same fight to save the Sound, the Gulf, the sand lances and the eels and the krill and the whales. In the Sound, whales almost entirely left the ecosystem in the 1960s and are only now returning to the western and central Sound after decades of legislation and thousands of volunteer and scientific and conservation hours. I had to build on what I knew in the Badlands of South Dakota, what I knew the first time I took Dex to Monhegan, what I knew the day my father had me stand on the cliff and look straight out to the sea, what I knew when I chose to have Dex—my life is worth saving.

Learning to live with generational loss is a muscle, like any other, that can be strengthened. We can use this muscle to spiral in and out of mourning, visit with the dead and grieve them, visit with the dead and welcome them at the table, to the river of creatures through the house, to the wedding, to the creek you can see while you say your vows, to the trilobite fossil around your neck.

To reckon with the whirlpool, you must learn that your life is powerful, no matter what vulnerabilities it entails. To learn to swim in the world as it is, you must learn how to lay the dead to rest and still need them as much as the living. To learn to mourn joyfully, you must learn to hold wonder and terror braided in your hand and not look away. To learn to live as a survivor, you must learn that you are not a failure because you are alive, and they are not. To learn to raise the next generation, you must believe that your children are not failures because they are imperfect.

In footage filmed by a fisherman in 2020, humpback whales leap from the Sound. I have watched it over and over again. They are not whales that stand in for the devil. They are not whales that reveal themselves to be an island. They are whales that had always been waiting to take me deeper, into the profound.

In the Monhegan harbor, we said yes to joy, to risk, whether we were mourning or not. "Go in, Mom! Get all the way in!"

I held my nose and dunked down into the water. It was soothing and I came up into the air and wind, staying low in the water, knees bent so that my shoulders were still in the cold, acclimating me. I could see the dock from there, the place where we met the squid, the place where the cutter picked us up that night. Dex swam up to me. "Better?"

"Better! Let's go farther out."

I turned to face the sunset. The red ball was moving to the lip of the world, widening and flattening, about to disappear behind Manana; the clouds stacked up over us, another flash of lightning hit the distance. "Wow."

The Earth is a miracle, and you are a miracle, and life is a miracle. Keep moving—into the depths of the ocean, onto the cliff, out to the star-scattered sky—to wherever you feel the terrifying whirlpool resolve into magnitude, magnificence, and the unknown. In that mystery lies all that can save you and the world entire. It is a terror. It is a wonder. It is a vortex. It is a loss. It is a healing. It is a spiral. It is a gift. It is ours. We have all we need to meet it, over and over again, with bravery that comes from a world made so perfectly that we are always accompanied, human to animal, tree to insect, fungus to plankton, tide pool to star, blood to water.

I kept walking forward, my knees straightening as I went deeper. The storm was over the Gulf, the sunset glow all through the harbor waters, the dock just to the east of us, where journeys start and end. Dex walked at my side. He was taller than me now, his long hair wet and waving. He took my hand under the water.

"Farther," he said. And we walked to where we could no longer touch the bottom.

And then we swam.

The End

Thank You

To Dex for the periwinkle song and the new legacy and the hope. To Andy, who said all the time, *You are the argument.*

To Julia Lord for being my agent and reading this not far from the city where I was born and the hills where I grew up—the Housatonic connects us. To Gina Frangello for friendship that spanned our illnesses and for buying the book and believing that I could dig deeper and risk more. To Sarah Leavitt, who tried to help me find it. To Eryn Loeb, who edited the chapter that made me understand. To Row House for taking the risk. To Martha Bayne for being an editor's editor and the illuminator of my beating heart. To Scott Foley and Jeffery Johnson for the movie. To Esther Bergdahl for fact-checking. To Daliah Saper for legal support. To the Millay Colony, Hedgebrook, the Banff Centre, and Ragdale, where I wrote this book, and to the broken kneecap residency where I wrote it all over again. To everyone at the Whiting Foundation for the honor and for funding the last great push for this book to come into the world as its best self.

To the people and creatures of Monhegan and its watershed, who shared their home with me, and to so many lovers of the place, especially Cat and Daphne, Mel and Stew, Jim and Jen, and the emergency workers. To the Coast Guard and Lincoln County

Hospital. To the hospices that cared for my parents. To the doctors and healers. To the people and creatures of the Long Island Sound and its watershed. To the scientists and activists and poets holding on to hope and fighting.

To Aunt June and Uncle Kurt and June's kids and everyone who cared for my mother and father—Sarah Whalen, Freddie Manning, Nancy Braxton, Laurie Brown, Tommy Thomas, and Jack and Duffy. To everyone who cared for me—Amanda, Bob and Judy, Molly and Keith, Karen and Tony, Shana and Simo, Marsha, Sherry and Dan, Daphne and Robert, Kelly and Airan, Taryn, Laurie Lee and Katy, Ian and Sam, Rosie, Walt, Bathsheba, Deidre, Libby, Carolyn and Felipe, Bill and Eric, Don and Gary, Seth and Susan, Lindsay J and Steve, my students at TLL, the Bolles family, and Spencer and the kids, the Families Together book club, and everyone in 6637.

To the artists who supported this—Lawrence Peters, Robin Cline, Amy Eaton, Kate Blomquist, Scott Ramsayer, Molia Dumbleton, Ariel Bolles, Zachariah Parker, Lindsay Porter, Dayna Bateman, Josephine Ferorelli, Tiffany Higgins, Tom Doig, Tony Birch, Eric Sasson, Gretchen Lida, Bethany Schneider, Jacqui Shine, Mairead Case, Maia Morgan, Tom Shachtman, Nance Klehm, Megan Stielstra, Jacob Hand, Karen Grigsby-Bates, Desiree Cooper, Rebecca Altman, Kerri Arsenault, Kate Olsen, Jennifer Lunden, Naomi Klein, Nayonmi Munaweera, Becky Brown, Sage Reed, The Otters, Zoe Zolbrod, Joshua Dumas, Sheila Sachs, Rob Roberge, and the people of Women & Children First Bookstore.

To Aunt Lois and Uncle Brian and Uncle Patrick, and my grandfather and great-grandfather, who I never met, for enduring the unendurable and leaving me a genetic legacy to both wrestle with and glory in. For you, and for all the people with PKD and

other chronic illnesses—I wish for some measure of the lost brilliance that was taken from you to find its way into the world.

To everyone who was a parachute when I felt pulled down by
an anchor and who was an anchor when I thought I would blow
away.

To my father for shaping me. To my mother for saving me.

A Note on the Text

This book is a researched work of nonfiction science as well as a memoir and has been fact-checked and undergone a legal read. To write it, I conducted many interviews and research trips to locations mentioned. Many of the people in this book passed away decades ago, and some stories and memories have been reconstructed through additional interviews and research. If my parents and my father's brothers and sisters were available to corroborate, these stories would also have been personally confirmed. Any errors of memory or research, therefore, are mine. Any living people who are named accurately here were given the chance to read and review and correct and veto the material in this book. Any living people who were not willing or able to read and review had their names and details changed to protect their anonymity. My child had full access to the text and full power to approve or deny the inclusion of any part of their story in the book. In addition, they had full access to multiple kinds of support in preparation for this story being told.

References

BOOKS

Andersen, Tom. 2002. *This Fine Piece of Water: An Environmental History of Long Island Sound.* Yale University Press.

Carson, Rachel. 1955. *The Edge of the Sea.* Houghton Mifflin Harcourt.

———. 1958. *The Sea Around Us.* Oxford University Press.

———. 1965. *The Sense of Wonder.* HarperCollins.

Cronon, William. 2011. *Changes in the Land: Indians, Colonists, and the Ecology of New England.* Hill and Wang.

Faller, Ruth Grant. 2001. *Monhegan: Her Houses and Her People, 1780–2000.* Mainstay Publications.

Goodridge, Harry, and Lew Dietz. 2014. *A Seal Called Andre.* Down East Books.

Harris, Lillian. 2014. *Monhegan Nature Guide: Natural History and Guided Hikes on One of Maine's Wildest Offshore Islands.* Yellow House Publishing.

Lipman, Andrew. 2015. *The Saltwater Frontier: Indians and the Contest for the American Coast.* Yale University Press.

Lynch, Patrick J. 2017. *A Field Guide to Long Island Sound: Coastal Habitats, Plant Life, Fish, Seabirds, Marine Mammals, and Other Wildlife.* Yale University Press.

REFERENCES BY CHAPTER

Longfin Inshore Squid

Alam, Ahsan, Émilie Cornec-Le Gall, and Ronald D. Perrone. 2023. "What Is Autosomal Dominant Polycystic Kidney Disease?" *JAMA* 329 (13): 1128. https://doi.org/10.1001/jama.2023.2161.

Balat, Ayşe. 2016. "Tear Drops of Kidney: A historical overview of Polycystic Kidney Disease." *Giornale Italiano Di Nefrologia* 33 Suppl 66 (February): 33.S66.21. https://pubmed.ncbi.nlm.nih.gov/26913889.

Bever, Fred. 2021. "Scientists Think Longfin Squid Are Reason for Maine's Prized Shrimp Fishery Collapse." *WBUR*, October 4, 2021. https://www.wbur.org/news/2021/10/04/maine-shrimp-longfin-squid-warming-waters.

Cordido, Adrián, Lara Besada-Cerecedo, and Miguel A. García-González. 2017. "The Genetic and Cellular Basis of Autosomal Dominant Polycystic Kidney Disease—A Primer for Clinicians." *Frontiers in Pediatrics* 5 (December). https://doi.org/10.3389/fped.2017.00279.

Crawford, Karen, Juan Felipe Diaz Quiroz, Kristen M. Koenig, Namrata Ahuja, Caroline B. Albertin, and Joshua J. C. Rosenthal. 2020. "Highly Efficient Knockout of a Squid Pigmentation Gene." *Current Biology* 30 (17): 3484–3490. e4. https://doi.org/10.1016/j.cub.2020.06.099.

Derby, Charles D. 2022. "Cephalopods and Neuroscience Go Arm in Arm in The Biological Bulletin." The Biological Bulletin 243 (1): 5–8. https://doi.org/10.1086/721792.

Dinneen, James. 2021. "Untangling Mysteries of the Brain—with the Remarkable Biology of Squid." *National Geographic*, July 27, 2021. https://www.nationalgeographic.com/science/article/untangling-mysteries-of-the-brain-with-the-remarkable-biology-of-squid.

Geller, Markham J., and Simon L. Cohen. 1995. "Kidney and Urinary Tract Disease in Ancient Babylonia, with Translations of the Cuneiform Sources." *Kidney International* 47 (6): 1811–15. https://doi.org/10.1038/ki.1995.250.

Mills, Kathy, Dave Reidmiller, and Adam Kemberling. 2023. "Gulf of Maine Warming Update: 2022 the Second-Hottest Year on Record." Gulf of Maine Research Institute. February 15, 2023. https://www.gmri.org/stories/warming-22/.

Napoli, Francesca, Christina M. Daly, Stephanie Neal, Kyle J. McCulloch, Alexandra R. Zaloga, Alicia Liu, and Kristen M. Koenig. 2022. "Cephalopod Retinal Development Shows Vertebrate-like Mechanisms of Neurogenesis." *Current Biology* 32 (23): 5045–5056.e3. https://doi.org/10.1016/j.cub.2022.10.027.

"Polycystic Kidney Disease: Overview." n.d. Mayo Clinic. Accessed December 27, 2023. https://www.mayoclinic.org/diseases-conditions/polycystic-kidney-disease/symptoms-causes/syc-20352820.

Poppick, Laura. 2018. "Why Is the Gulf of Maine Warming Faster than 99% of the Ocean?" *Eos*, November 12, 2018. https://eos.org/features/why-is-the-gulf-of-maine-warming-faster-than-99-of-the-ocean.

Schwiening, Christof J. 2012. "A Brief Historical Perspective: Hodgkin and Huxley." *The Journal of Physiology* 590 (11): 2571–75. https://doi.org/10.1113/jphysiol.2012.230458.

"What Is a Bestiary?" n.d. The Aberdeen Bestiary. https://www.abdn.ac.uk/bestiary/what.php.

Wu, Min, Diping Wang, Ladan Zand, Peter C. Harris, Wendy White, Vesna D. Garovic, and Cindy A. Kermott. 2015. "Pregnancy Outcomes in Autosomal Dominant Polycystic Kidney Disease: A case-control study." *Journal of Maternal-Fetal & Neonatal Medicine* 29 (5): 807–12. https://doi.org/10.3109/14767058.2015.1019458.

Longneck Clam

Alexander, Shana. 1962. "They Decide Who Lives, Who Dies: Medical miracle and a moral burden of a small committee." *LIFE*. November 9, 1962; 53(19):102–4, 106, 108, 110, 115, 117–8, 123–24. http://books.google.com/books?id=qUoEAAAAMBAJ.

Blagg, Christopher R. 2007. "The Early History of Dialysis for Chronic Renal Failure in the United States: A View from Seattle." *American Journal of Kidney Diseases* 49 (3): 482–96. https://doi.org/10.1053/j.ajkd.2007.01.017.

Butler, Catherine R., Rajnish Mehrotra, Mark R. Tonelli, and Daniel Lam. 2016. "The Evolving Ethics of Dialysis in the United States: A Principlist Bioethics Approach." *Clinical Journal of the American Society of Nephrology* 11 (4): 704–9. https://doi.org/10.2215/cjn.04780515.

Cassidy, Emily, and NASA Earth Observatory. 2022. "Breathing Life into the Ocean." December 3, 2022. https://earthobservatory.nasa.gov/images/150656/breathing-life-into-the-ocean.

Cusick, Kathleen D., and Gary S. Sayler. 2013. "An Overview on the Marine Neurotoxin, Saxitoxin: Genetics, Molecular Targets, Methods of Detection and Ecological Functions." *Marine Drugs* 11 (4): 991–1018. https://doi.org/10.3390/md11040991.

Feder, Bernard. 1975. "When the Red Tide Strikes." *New York Times*, August 24, 1975. https://www.nytimes.com/1975/08/24/archives/when-the-red-tide-strikes-when-the-red-tide-strikes.html.

Fields, Robin. 2010. "Dialysis: An Experiment in Universal Health Care." *Fresh Air*, November 9, 2010. Interview by Terry Gross. NPR. https://www.npr.org /transcripts/131167638.

Gutis, Philip S. 1988. "Troubled Seas: Global Red Tides of Algae Bring New Fears." *New York Times*, May 3, 1988. https://www.nytimes.com/1988/05/03 /science/troubled-seas-global-red-tides-of-algae-bring-new-fears.html.

Institute for Quality and Efficiency in Health Care (IQWiG). 2018. "How Does Dialysis Work?" InformedHealth.Org. March 8, 2018. https://www.ncbi.nlm .nih.gov/books/NBK492981/.

Lindsey, Rebecca, and Michon Scott. 2010. "What Are Phytoplankton?" NASA Earth Observatory. July 13, 2010. https://earthobservatory.nasa.gov/features /Phytoplankton.

Manjoney, Delia Mary, and F. Patrick McKegney. 1979. "Individual and Family Coping with Polycystic Kidney Disease: The Harvest of Denial." *The International Journal of Psychiatry in Medicine* 9 (1): 19–31. https://doi.org/10.2190/nn3e -5ke8-k2r0-la3l.

Martin, Jennifer L., Murielle M. LeGresley, and Alex Hanke. 2014. "Thirty years—*Alexandrium fundyense* Cyst, Bloom Dynamics and Shellfish Toxicity in the Bay of Fundy, Eastern Canada." *Deep Sea Research Part II: Topical Studies in Oceanography* 103 (May): 27–39. https://doi.org/10.1016/j.dsr2.2013.08.004.

McGillicuddy, Dennis J., Michael L. Brosnahan, Darcie A. Couture, Ruoying He, Bruce A. Keafer, James P. Manning, Jennifer L. Martin, Cynthia H. Pilskaln, David W. Townsend, and Donald M. Anderson. 2014. "A Red Tide of *Alexandrium fundyense* in the Gulf of Maine." *Deep Sea Research Part II: Topical Studies in Oceanography* 103 (May): 174–84. https://doi.org/10.1016/j .dsr2.2013.05.011.

Mitchell, Jennifer. 2019. "Testing the Ocean, One Teaspoon at a Time." *Maine Public*, December 15, 2019. https://www.mainepublic.org/environment-and -outdoors/2019-12-15/testing-the-ocean-one-teaspoon-at-a-time.

Rettig, Richard A. 1991. "Origins of the Medicare Kidney Disease Entitlement: The Social Security Amendments of 1972." *Biomedical Politics*. 1991. National Academies Press (US). 1991. https://www.ncbi.nlm.nih.gov/books/NBK234191/.

"Nutrients and Eutrophication." 2019. U.S. Geological Survey. March 3, 2019. https://www.usgs.gov/mission-areas/water-resources/science/nutrients-and -eutrophication.

Professional, Cleveland Clinic Medical. 2021. "Dialysis." Cleveland Clinic. https://my.clevelandclinic.org/health/treatments/14618-dialysis.

Ross, Will. 2012. "God Panels and the History of Hemodialysis in America: A Cautionary Tale." *AMA Journal of Ethics* 14 (11): 890–96. https://doi.org/10.1001/virtualmentor.2012.14.11.mhst1-1211.

TIME. 1969. "America's Sewage System and the Price of Optimism," August 1, 1969. https://content.time.com/time/subscriber/article/0,33009,901182,00.html.

Long-Armed Hermit Crab

Anonymous. 1998. "Innovations: The History of Brass Making in the Naugatuck Valley." Copper Development Association. March 1998. https://www.copper.org/publications/newsletters/innovations/1998/03/naugatuck.html.

Casey, Ann C., Brian Bush, and David O. Carpenter. 2022. "PCBs in Indoor Air and Human Blood in Pittsfield, Massachusetts." *Chemosphere* 293 (April): 133551. https://doi.org/10.1016/j.chemosphere.2022.133551.

Long Island Sound Study. 2023a. "By the Numbers." https://longislandsoundstudy.net/about-the-sound/by-the-numbers/.

———. 2023b. "Water Quality Monitoring in the Sound and Embayments." June 27, 2023. https://longislandsoundstudy.net/research-monitoring/water-quality-monitoring/.

Manjoney, Delia Mary, and F. Patrick McKegney. 1979. "Individual and Family Coping with Polycystic Kidney Disease: The Harvest of Denial." *The International Journal of Psychiatry in Medicine* 9 (1): 19–31. https://doi.org/10.2190/nn3e-5ke8-k2r0-la3l.

Rossos, Katrina M., Tomas W. Smieszek, Jeffrey R. Barbaro, and New England Water Science Center. 2023. "Long Island Sound Overview." U.S. Geological Survey. March 7, 2023. https://www.usgs.gov/tools/overview-long-island-sound.

"Understanding PCB Risks at the GE-Pittsfield/Housatonic River Site." 2023. U.S. Environmental Protection Agency. https://www.epa.gov/ge-housatonic/understanding-pcb-risks-ge-pittsfieldhousatonic-river-site.

Wajda, Shirley T., PhD. 2023. "Ending the Danbury Shakes: A Story of Workers' Rights and Corporate Responsibility." *Connecticut History* (blog). September 26, 2023. https://connecticuthistory.org/ending-the-danbury-shakes-a-story-of-workers-rights-and-corporate-responsibility/.

Atlantic Slipper Shell

Blanchard, Michel. 2009. "Crepidula fornicata (American slipper limpet)." Data set. *CABI Compendium*. https://doi.org/10.1079/cabicompendium.108234.

Chan, Kit Yu Karen, Houshuo Jiang, and Dianna K. Padilla. 2013. "Swimming Speed of Larval Snail Does Not Correlate with Size and Ciliary Beat Frequency." *PLOS ONE* 8 (12): e82764. https://doi.org/10.1371/journal.pone.0082764.

Chan, Ramony, Robert T. Brooks, Jonathan Erlich, Josephine Chow, and Michael Suranyi. 2009. "The Effects of Kidney-Disease-Related Loss on Long-Term Dialysis Patients' Depression and Quality of Life." Clinical Journal of the American Society of Nephrology 4 (1): 160–67. https://doi.org/10.2215/cjn.01520308.

Kaufman, Kenn. n.d. "Sanderling." Audubon Bird Guide. Accessed December 27, 2023. https://www.audubon.org/field-guide/bird/sanderling. Adapted from Kenn Kaufman, Lives of North American Birds (1996, Houghton Mifflin Harcourt).

Kelley, George C. 1977. "Late Pleistocene and Recent Geology of the Housatonic River Region in Northwestern Connecticut." USGS Publications Warehouse. https://doi.org/10.3133/ofr77545.

Moreno-Parada, Daniel, Jorge Álvarez-Solas, Javier Blasco, Marisa Montoya, and Alexander Robinson. 2023. "Simulating the Laurentide Ice Sheet of the Last Glacial Maximum." *The Cryosphere* 17 (5): 2139–56. https://doi.org/10.5194/tc-17-2139-2023.

"New England Native American Groups." n.d. National Geographic Education. Accessed December 27, 2023. https://education.nationalgeographic.org/resource/new-england-native-american-groups/.

Smith, Paul A., Adam C. Smith, Brad A. Andres, Charles M. Francis, Brian Harrington, Christian Friis, R. I. G. Morrison, Julie Paquet, Brad Winn, and Stephen C. Brown. 2023. "Accelerating Declines of North America's Shorebirds Signal the Need for Urgent Conservation Action." *Ornithological Applications* 125 (2). https://doi.org/10.1093/ornithapp/duad003.

Swiniarski, Ava, Jonah Hollis, and Ben Smith. 2017. "Dam . . . the Atlantic Salmon Are Gone." *Debating Science* (blog). December 5, 2017. https://blogs.umass.edu/natsci397a-eross/dam-the-atlantic-salmon-are-gone/.

Tong, Allison, Gopala K. Rangan, Marinella Ruospo, Valeria Saglimbene, Giovanni F.M. Strippoli, Suetonia C. Palmer, David J. Tunnicliffe, and Jonathan

C. Craig. 2015. "A Painful Inheritance—Patient Perspectives on Living with Polycystic Kidney Disease: Thematic Synthesis of Qualitative Research." *Nephrology Dialysis Transplantation* 30 (5): 790–800. https://doi.org/10.1093/ndt/gfv010.

Viard, Frédérique, Céline Ellien, and Lise Dupont. 2006. "Dispersal Ability and Invasion Success of Crepidula Fornicata in a Single Gulf: Insights from Genetic Markers and Larval-Dispersal Model." *Helgoland Marine Research* 60 (2): 144–52. https://doi.org/10.1007/s10152-006-0033-8.

Blood Ark

Benjamin, Vernon. 2018. "Whaling on the Hudson." Hudson River Maritime Museum. June 15, 2018. https://www.hrmm.org/history-blog/whaling-on-the-hudson.

Blagg, Christopher. 1996. "A Brief History of Home Hemodialysis." *Advances in Renal Replacement Therapy* 3 (2): 99–105. https://doi.org/10.1016/s1073-4449(96)80048-3.

Fenoff, Linda. n.d. "History." City of Hudson, New York. Accessed December 27, 2023. https://www.hudsonny.gov/visitors/history/index.php.

United States Renal Data System. 2022 *USRDS Annual Data Report: Epidemiology of kidney disease in the United States.* "End Stage Renal Disease: Chapter 2: Home Dialysis." National Institutes of Health, National Institute of Diabetes and Digestive and Kidney Diseases. https://usrds-adr.niddk.nih.gov/2022/end-stage-renal-disease/2-home-dialysis.

Jingle Shell

Ashton, Gail, Amy L. Freestone, J. Emmett Duffy, Mark E. Torchin, Brent J. Sewall, Brianna M. Tracy, Mariano Albano, et al. 2022. "Predator Control of Marine Communities Increases with Temperature across 115 Degrees of Latitude." *Science* 376 (6598): 1215–19. https://doi.org/10.1126/science.abc4916.

Bansal, Rahul, and Anil Kapoor. 2014. "Laparoscopic Nephrectomy for Massive Polycystic Kidney Disease: Updated Technique and Outcomes." *Canadian Urological Association Journal* 8 (9–10): 341. https://doi.org/10.5489/cuaj.2097.

Becker, Mary, and Water Monitoring and Assessment Program. 2023. "Long Island Sound Water Quality and Hypoxia Monitoring Program Overview." CT.Gov. https://portal.ct.gov/DEEP/Water/LIS-Monitoring LIS-Water-Quality-and-Hypoxia-Monitoring-Program-Overview.

Bhutta, Beenish S., Faysal Alghoula, and Ilya Berim. 2022. "Hypoxia." StatPearls. August 9, 2022. https://www.ncbi.nlm.nih.gov/books/NBK482316/.

Casey, Michelle, Gregory P. Dietl, David M. Post, and Derek E. G. Briggs. 2014. "The Impact of Eutrophication and Commercial Fishing on Molluscan Communities in Long Island Sound, USA." *Biological Conservation* 170 (February): 137–44. https://doi.org/10.1016/j.biocon.2013.12.037.

Cheng, Brian S., Lisa M. Komoroske, and Edwin D. Grosholz. 2016. "Trophic Sensitivity of Invasive Predator and Native Prey Interactions: Integrating Environmental Context and Climate Change." *Functional Ecology* 31 (3): 642–52. https://doi.org/10.1111/1365-2435.12759.

D'Entremont, Jeremy. n.d. "History of Penfield Reef Lighthouse, Fairfield, Connecticut." New England Lighthouses: A Virtual Guide. Accessed December 27, 2023. http://www.newenglandlighthouses.net/penfield-reef-light-history.html.

Long Island Sound Study. 2023. "Hypoxia." https://longislandsoundstudy.net /about/our-mission/management-plan/hypoxia/.

Lord, Joshua P., and Robert B. Whitlatch. 2013. "Impact of Temperature and Prey Shell Thickness on Feeding of the Oyster Drill *Urosalpinx cinerea* Say." *Journal of Experimental Marine Biology and Ecology* 448 (October): 321–26. https://doi.org/10.1016/j.jembe.2013.08.006.

Malmquist, David. n.d. "Dead Zones." Virginia Institute of Marine Science. Accessed December 27, 2023. https://www.vims.edu/research/topics/dead _zones/index.php.

Mississippi River/Gulf of Mexico Hypoxia Task Force. 2023. "Hypoxia 101." U.S. Environmental Protection Agency. https://www.epa.gov/ms-htf /hypoxia-101.

Nordham, Kristen D., and Scott Ninokawa. 2021. "The History of Organ Transplantation." *Baylor University Medical Center Proceedings* 35 (1): 124–28. https://doi.org/10.1080/08998280.2021.1985889.

Poletajew, Slawomir, Artur A. Antoniewicz, and Andrzej Borówka. 2010. "Kidney Removal: The past, presence, and perspectives: A historical review." *Urology Journal* 7 (4). https://pubmed.ncbi.nlm.nih.gov/21170847/.

Ravo, Nick. 1987. "A Portion of L.I. Sound Has No Oxygen or Fish." *New York Times*, July 31, 1987. https://www.nytimes.com/1987/07/31/nyregion/a-portion -of-li-sound-has-no-oxygen-or-fish.html.

Rettig, Richard A. 1991. "Origins of the Medicare Kidney Disease Entitlement: The Social Security Amendments of 1972." *Biomedical Politics*. https://www.ncbi .nlm.nih.gov/books/NBK234191/.

Vaudrey, Jamie. 2017. "New York City's Impact on Long Island Sound Water Quality Technical Report." Vaudrey Lab, Department of Marine Sciences, University of Connecticut. November 18, 2017. https://vaudrey.lab.uconn.edu /wp-content/uploads/sites/1663/2018/07/2017-11-16-Vaudrey-NYC-N.pdf.

Dinoflagellate Microalgae

Crane, Josh. 2020. "Angel's Glow: From Civil War Folklore to Winning Science Fair Project." WBUR. https://www.wbur.org/endlessthread/ 2020/08/21/angels -glow-science-fair.

Dybas, Cheryl. 2019. "Illuminating New Biomedical Discoveries: Bioluminescent, biofluorescent species glow with promise." *BioScience* 69 (7): 487–95. https://doi .org/10.1093/biosci/biz052.

Eck, Allison. 2018. "Bioluminescent Bays Shine Light on Puerto Rico's Resilience." *NOVA*, September 17, 2018. https://www.pbs.org/wgbh/nova/article /bioluminescent-bays-shine-light-on-puerto-ricos-resilience/.

Hanley, Karen A., and Edith A. Widder. 2017. "Bioluminescence in Dinoflagellates: Evidence that the Adaptive Value of Bioluminescence in Dinoflagellates Is Concentration Dependent." *Photochemistry and Photobiology* 93 (2): 519–30. https://doi.org/10.1111/php.12713.

Jabr, Ferris. 2016. "The Secret History of Bioluminescence." *Hakai Magazine*, May 10, 2016. https://hakaimagazine.com/features/secret-history-bioluminescence/.

Jokinen, Sami A., Joonas J. Virtasalo, Tom Jilbert, Jérôme Kaiser, Olaf Dellwig, Helge Wolfgang Arz, Jari Hänninen, Laura Arppe, Miia Collander, and Timo Saarinen. 2018. "A 1500-Year Multiproxy Record of Coastal Hypoxia from the Northern Baltic Sea Indicates Unprecedented Deoxygenation over the 20th Century." *Biogeosciences* 15 (13): 3975–4001. https://doi.org/10.5194/bg-15 -3975-2018.

Maksel, Rebecca. 2016. "Jim Lovell, from Carriers to the Moon." *Smithsonian Magazine*, April 15, 2016. https://www.smithsonianmag.com/air-space -magazine/in-the-museum-space-capsules-180958440/.

Nalewicki, Jennifer. 2022. "Puerto Rico's Bioluminescent Bays Are Brighter than Ever." *Smithsonian Magazine*, April 6, 2022. https://www.smithsonianmag.com /travel/puerto-ricos-bioluminescent-bays-are-brighter-than-ever-180979874/.

Nyholm, Spencer V., and Margaret J. McFall-Ngai. 2021. "A Lasting Symbiosis: How the Hawaiian Bobtail Squid Finds and Keeps Its Bioluminescent Bacterial Partner." *Nature Reviews Microbiology* 19 (10): 666–79. https://doi.org/10.1038 /s41579-021-00567-y.

Rameshkumar, P., P.S. Thirumalaiselvan, Mini Raman, L Remya, R Jayakumar, Mohammed Sakthivel, G Tamilmani, et al. 2023. "Monitoring of Harmful Algal Bloom (HAB) of Noctiluca Scintillans (Macartney) along the Gulf of Mannar, India Using in-Situ and Satellite Observations and Its Impact on Wild and Maricultured Finfishes." *Marine Pollution Bulletin* 188 (March): 114611. https:// doi.org/10.1016/j.marpolbul. 2023.114611.

Valiadi, Martha, and Debora Iglesias-Rodríguez. 2013. "Understanding Bioluminescence in Dinoflagellates—How Far Have We Come?" *Microorganisms* 1 (1): 3–25. https://doi.org/10.3390/microorganisms1010003.

Williams, Catrin F. 2018. "What Is Bioluminescence and How Is It Used by Humans and in Nature?" *The Conversation*, July 27, 2018. https://theconversation. com/what-is-bioluminescence-and-how-is-it-used-by-humans-and-in -nature-100472.

Northern Rock Barnacle

Gillis, Justin. 2017. "The Sea Level Did, in Fact, Rise Faster in the Southeast U.S." *New York Times*, August 10, 2017. https://www.nytimes.com/2017/08/09 /climate/the-sea-level-did-in-fact-rise-faster-in-the-southeast-us.html.

Mooney, Chris, and Brady Dennis. 2023. "Seas Have Drastically Risen along Southern U.S. Coast in Past Decade." *Washington Post*, April 1, 2023. https:// www.washingtonpost.com/climate-environment/ 2023/04/10/sea-level-rise -southern-us/.

Nunez, Joaquin C. B., Patrick A. Flight, Kimberly B. Neil, Stephen Rong, Leif A. Eriksson, David A. Ferranti, Magnus Alm Rosenblad, Anders Blomberg, and David M. Rand. 2020. "Footprints of Natural Selection at the Mannose-6 -Phosphate Isomerase Locus in Barnacles." *Proceedings of the National Academy of Sciences of the United States of America* 117 (10): 5376–85. https://doi.org /10.1073/pnas.1918232117.

Polyakov, I. S., Randi Ingvaldsen, Andrey V. Pnyushkov, Uma S. Bhatt, Jennifer A. Francis, Markus Janout, R. Kwok, and Øystein Skagseth. 2023. "Fluctuating Atlantic Inflows Modulate Arctic Atlantification." *Science* 381 (6661): 972–79. https://doi.org/10.1126/science.adh5158.

Rantanen, Mika, Alexey Yu. Karpechko, Antti Lipponen, Kalle Nordling, Otto Hyvärinen, Kimmo Ruosteenoja, Timo Vihma, and A. Laaksonen. 2022. "The Arctic Has Warmed Nearly Four Times Faster than the Globe since 1979." *Communications Earth & Environment* 3 (1). https://doi.org/10.1038/s43247-022 -00498-3.

Reidmiller, Dave. 2023. "Causes & Impacts of Rapid Warming." *Gulf of Maine, Explained* (blog). August 17, 2023. https://www.gmri.org/stories/gulf-of-maine -explained-causes-impacts-of-rapid-warming/.

Sayers, Paul, Charlotte Moss, Sam Carr, and Andrés Payo. 2022. "Responding to Climate Change around England's Coast - The Scale of the Transformational Challenge." *Ocean & Coastal Management* 225 (June): 106187. https://doi .org/10.1016/j.ocecoaman.2022.106187.

Stevenson, David K., Susan Tuxbury, Michael R. Johnson, and Christopher Boelke. 2014. "Shallow Water Benthic Habitats in the Gulf of Maine: A Summary of Habitat Use by Common Fish and Shellfish Species in the Gulf of Maine." *Greater Atlantic Region Policy Series - NOAA Fisheries*, May. https://www .greateratlantic.fisheries.noaa.gov/policyseries/index.php/GARPS/article/view/4.

Sundell, Kristina, Anna-Lisa Wrange, Per R. Jonsson, and Anders Blomberg. 2019. "Osmoregulation in Barnacles: An Evolutionary Perspective of Potential Mechanisms and Future Research Directions." *Frontiers in Physiology* 10 (August). https://doi.org/10.3389/fphys.2019.00877.

Thompson, Colleen and The Gulf of Maine Council on the Marine Environment. 2010. "The Gulf of Maine in Context: State of the Gulf of Maine Report." http:// www.gulfofmaine.org/state-of-the-gulf/docs/the-gulf-of-maine-in-context.pdf.

Copepod

Bigelow Laboratory for Ocean Sciences. 2016. "Water Color and Phytoplankton Growth in the Gulf of Maine Are Changing." *Phys.Org*, April 21, 2016. https:// phys.org/news/2016-04-phytoplankton-growth-gulf-maine.html.

Goździejewska, Anna, and Marek Kruk. 2023. "The Response of Zooplankton Network Indicators to Winter Water Warming Using Shallow Artificial

Reservoirs as Model Case Study." *Scientific Reports* 13 (1). https://doi
.org/10.1038/s41598-023-45430-7.

Jackson, Derrick Z. Jackson. February 20, 2022. "A Big Climate Warning from
One of the Gulf of Maine's Smallest Marine Creatures." *Inside Climate News*,
February 18, 2022. https://insideclimatenews.org/news/20022022/maine
-marine-ecosystem-calanus-finmarchicus-climate-change/.

Lloyd, Jared. 2016. "The Gulf Stream and the Age of Exploration." *Coastal
Review* (blog). April 1, 2016. https://coastalreview.org/2016/04/13740/.

Nobre, Paulo, Sandro F. Veiga, Emanuel Giarolla, André Lanfer Marquez,
Manoel B. Da Silva, Vinícius Capistrano, Marta Malagutti, et al. 2023. "AMOC
Decline and Recovery in a Warmer Climate." *Scientific Reports* 13 (1). https://
doi.org/10.1038/s41598-023-43143-5.

Ohman, Mark D., and Jefrey A. Runge. 1994. "Sustained Fecundity When
Phytoplankton Resources Are in Short Supply: Omnivory by Calanus
Finmarchicus in the Gulf of St. Lawrence." *Limnology and Oceanography* 39 (1):
21–36. https://doi.org/10.4319/lo.1994.39.1.0021.

Rahmstorf, Stefan. 2023. "What Is Happening in the Atlantic Ocean to the
AMOC?" *RealClimate* (blog). July 26, 2023. https://www.realclimate.org/index
.php/archives/2023/07/what-is-happening-in-the-atlantic-ocean-to-the-amoc/.

Richardson, Philip L. 1980. "The Benjamin Franklin and Timothy Folger Charts
of the Gulf Stream." *Third International Congress on the History of Oceanography*,
September. https://www2.whoi.edu/staff/prichardson/wp-content/uploads
/sites/75/2018/11/Richardson-1980-The-Benjamin-Franklin-and-Timothy
-Folger-charts-of-the-Gulf-Stream.pdf.

Rigby, Susan, and Clare V. Milsom. "Origins, Evolution, and Diversification of
Zooplankton." *Annual Review of Ecology and Systematics* 31 (2000): 293–313.
http://www.jstor.org/stable/221734.

Roncalli, Vittoria, Jefferson T. Turner, David M. Kulis, Donald M. Anderson,
and Petra H. Lenz. 2016. "The Effect of the Toxic Dinoflagellate Alexandrium
Fundyense on the Fitness of the Calanoid Copepod Calanus Finmarchicus."
Harmful Algae 51 (January): 56–66. https://doi.org/10.1016/j.hal.2015.11.003.

Sun, Jing, Jonathan D. Todd, J. Cameron Thrash, Yanping Qian, Michael C.
Qian, Ben Temperton, Jiazhen Guo, et al. 2016. "The Abundant Marine
Bacterium Pelagibacter Simultaneously Catabolizes Dimethylsulfoniopropionate

to the Gases Dimethyl Sulfide and Methanethiol." *Nature Microbiology* 1 (8). https://doi.org/10.1038/nmicrobiol.2016.65.

Weijer, Wilbert, Wei Cheng, Sybren Drijfhout, Alexey V. Fedorov, Aixue Hu, Laura Jackson, W. Liu, Elaine L. McDonagh, Jenny Mecking, and J. Zhang. 2019. "Stability of the Atlantic Meridional Overturning Circulation: A Review and Synthesis." *Journal of Geophysical Research: Oceans* 124 (8): 5336–75. https://doi.org/10.1029/2019jc015083.

American Lobster

Acharya, Aninda B. 2023. "Anosognosia." StatPearls. April 24, 2023. https://www.ncbi.nlm.nih.gov/books/NBK513361/.

"American Lobster Settlement Index." 2023. The Wahle Lab. July 28, 2023. https://umaine.edu/wahlelab/american-lobster-settlement-index-alsi/.

Ascher, Alex. 2020. "On the Appetites of Baby Lobsters." Maine Sea Grant. September 22, 2020. https://seagrant.umaine.edu/2020/09/22/on-the-appetites-of-baby-lobsters/.

Bever, Fred. 2018. "New Research Shows 'Strong Correlation' between Baby Lobster Decline, Possible Food Source." *Maine Public*, March 27, 2018. https://www.mainepublic.org/environment-and-outdoors/2018-03-27/new-research-shows-strong-correlation-between-baby-lobster-decline-possible-food-source.

———. 2018. "Research Concludes Maine Conservation Technique Helped Drive Lobster Population Boom." *Maine Public*, January 22, 2018. https://www.mainepublic.org/news/2018-01-22/research-concludes-maine-conservation-technique-helped-drive-lobster-population-boom.

Bris, Arnault Le, Katherine E. Mills, Richard A. Wahle, Yong Chen, Michael A. Alexander, Andrew Allyn, Justin G. Schuetz, James D. Scott, and Andrew J. Pershing. 2018. "Climate Vulnerability and Resilience in the Most Valuable North American Fishery." *Proceedings of the National Academy of Sciences of the United States of America* 115 (8): 1831–36. https://doi.org/10.1073/pnas.1711122115.

Carloni, Joshua T., Richard A. Wahle, Paul Geoghegan, and Eric P. Bjorkstedt. 2018. "Bridging the Spawner-Recruit Disconnect: Trends in American Lobster Recruitment Linked to the Pelagic Food Web." *Bulletin of Marine Science* 94 (3): 719–35. https://doi.org/10.5343/bms.2017.1150.

Crossin, Glenn T., S Al-Ayoub, Sh Jury, W. Huntting Howell, and Winsor H. Watson. 1998. "Behavioral Thermoregulation in the American Lobster Homarus Americanus." *The Journal of Experimental Biology* 201 (3): 365–74. https://doi .org/10.1242/jeb.201.3.365.

Hannachi, Abdel, and Waheed Iqbal. 2019. "On the Nonlinearity of Winter Northern Hemisphere Atmospheric Variability." *Journal of the Atmospheric Sciences* 76 (1): 333–56. https://doi.org/10.1175/jas-d-18-0182.1.

Maine Department of Marine Resources. 2023. "Historical Maine Lobster Landings." Data set. *Historical Maine Fisheries Landings Data.* Maine.gov. https:// www.maine.gov/dmr/fisheries/commercial/landings-program/historical-data.

———. n.d. "DMR Boothbay Harbor Environmental Monitoring Program." Maine.Gov. Accessed December 29, 2023. https://www.maine.gov/dmr/science /weather-tides/boothbay-harbor-environmental-data.

Nargi, Lela. 2021. "An Uptick in Industrial Aquaculture in Maine Has Some Lobster- and Fishermen Hot under the Collar." *The Counter*, August 5, 2021. https://thecounter.org/uptick-industrial-aquaculture-maine-lobster-fishermen/.

Reardon, Kathleen. 2012. "So You Want to Know: Why Do We V-Notch Lobsters?" Maine Lobstermen's Community Alliance. January 17, 2012.

Seara, Tarsila, Adrien Owens, Richard B. Pollnac, Robert S. Pomeroy, and Christopher L. Dyer. 2022. "Lessons Learned from a Natural Resource Disaster: The Long-Term Impacts of the Long Island Sound Lobster Die-off on Individuals and Communities." *Marine Policy* 136 (February): 104943. https:// doi.org/10.1016/j.marpol.2021.104943.

Tamman, Maurice. 2018. "Ocean Shock: The Great Lobster Rush." *Reuters*, October 30, 2018. https://www.reuters.com/investigates/special-report/ocean -shock-lobster/.

"Title 12, §6471: Monhegan Lobster Conservation Area." n.d. https://www .mainelegislature.org/legis/statutes/12/title12sec6471.html.

Wallace, Ryan, Charles S. Colgan, Middlebury Institute of International Studies at Monterey Center for the Blue Economy, and Maine Center for Business and Economic Research, University of Maine. 2023. "The Economic Impacts of the Maine Seafood Sector: Prepared for Seafood Economic Accelerator for Maine." https://www.seamaine.org/wp-content/uploads/2023/03/FINAL-SEAMaine -Economic-Impact-Analysis-Report-2.pdf.

Whittle, Patrick. 2018. "Study: Conservation Has Helped Lobsters as Oceans Warm." *Associated Press*, January 22, 2018. https://apnews.com/general-news-b0d746e859b046eaa5825a5f02105b7c.

Brook Trout

Al-Mansour, Zeina, Beverly P. Nelson, and Andrew M. Evens. 2013. "Post-Transplant Lymphoproliferative Disease (PTLD): Risk Factors, Diagnosis, and Current Treatment Strategies." *Current Hematologic Malignancy Reports* 8 (3): 173–83. https://doi.org/10.1007/s11899-013-0162-5.

Poole, Robert M. 2007. "Native Trout Are Returning to America's Rivers." *Smithsonian Magazine*, November 19, 2013. https://www.smithsonianmag.com/science-nature/native-trout-are-returning-to-americas-rivers-160317057/.

Sherman, Annie. 2022. "The Centuries-Old Fish Ladder That Feeds Maine's Lobster Industry." *Atlas Obscura*, April 26, 2022. https://www.atlasobscura.com/articles/fish-ladder-maine-lobster-industry.

Tuff, Nicholas. "The Natural History of the Berkshires." The Berkshires. https://berkshires.com/guides/to-do-see-guides/the-natural-history-of-the-berkshires/.

Common Periwinkle

Benoiston, Anne-Sophie, Federico M. Ibarbalz, Lucie Bittner, Lionel Guidi, O. Jahn, Stephanie Dutkiewicz, and Chris Bowler. 2017. "The Evolution of Diatoms and Their Biogeochemical Functions." *Philosophical Transactions of the Royal Society B* 372 (1728): 20160397. https://doi.org/10.1098/rstb.2016.0397.

Broun, Stacy N. 1998. "Understanding 'Post-AIDS Survivor Syndrome': A Record of Personal Experiences." *Aids Patient Care and STDs* 12 (6): 481–88. https://doi.org/10.1089/apc.1998.12.481.

Carlton, James T., and Charles S. Elton. 1992. "Introduced Marine and Estuarine Mollusks of North America: An End-of-the-20th-Century Perspective." *Journal of Shellfish Research* 11 (2): 489–505. https://www.researchgate.net/publication/255642084_Introduced_marine_and_estuarine_mollusks_of_North_America_an_end-of-the-20th-century_perspective.

Chen, Anelise. 2022. "What It Would Take to See the World Completely Differently." *The Atlantic*, May 17, 2022. https://www.theatlantic.com/books/archive/2022/05/rachel-carson-book-sea-trilogy-wonder/629842/.

Heschel, Abraham Joshua. 1976. *Man Is Not Alone: A Philosophy of Religion.* Macmillan.

Huizenga, Kristin. 2018. "Holding Fast: Kelp in Nova Scotia Tries to Grow on Turf Algae after a Period of Decline." *Oceanbites* (blog). September 6, 2018. https://oceanbites.org/holding-fast-kelp-in-nova-scotia-tries-to-grow-on-turf-algae-after-a-period-of-decline/.

Malviya, Shruti, Eleonora Scalco, Stéphane Audic, Flora Vincent, Alaguraj Veluchamy, Julie Poulain, Patrick Wincker, et al. 2016. "Insights into Global Diatom Distribution and Diversity in the World's Ocean." *Proceedings of the National Academy of Sciences of the United States of America* 113 (11). https://doi.org/10.1073/pnas.1509523113.

Marine Invasion Lab and Smithsonian Environmental Research Center. n.d. "Littorina Littorea." Nemesis. Accessed December 29, 2023. https://invasions.si.edu/nemesis/species_summary/70419.

Materne, M., T. Bush, M. Houck, and S. Snell. "Plant Guide for Smooth Cordgrass (Spartina alterniflora)." USDA Natural Resources Conservation Service, Louisiana State Office. Baton Rouge, LA. "Smooth Cordgrass." 2022. https://plants.usda.gov/DocumentLibrary/plantguide/pdf/pg_spal.pdf.

Morris. 2019. "Thoughts and Reflections on James Agee and 'A Death in the Family.'" *Speaking Volumes* (blog). November 11, 2019. https://volumes.lib.utk.edu/features/thoughts-and-reflections-on-james-agee-and-a-death-in-the-family/.

Pike, Sue. 2016. "Three Kinds of Periwinkles Live in Gulf of Maine." *Portsmouth Herald*, July 19, 2016. https://www.seacoastonline.com/story/news/local/2016/07/19/three-kinds-periwinkles-live-in/27466916007/.

U.S. Department of Agriculture. n.d. "Zebra Mussel." National Invasive Species Information Center. Accessed December 29, 2023. https://www.invasivespeciesinfo.gov/aquatic/invertebrates/zebra-mussel.

Zonana, Victor F. 1989. "Survivor's Syndrome: AIDS Takes Toll on Ones Left Behind." *Los Angeles Times*, May 6, 1989. https://www.latimes.com/archives/la-xpm-1989-05-06-mn-2084-story.html.

Comb Jellyfish

Beman, J. Michael, Kevin R. Arrigo, and Pamela A. Matson. 2005. "Agricultural Runoff Fuels Large Phytoplankton Blooms in Vulnerable Areas of the Ocean." *Nature* 434 (7030): 211–14. https://www.nature.com/articles/nature03370.

Fields, Scott. 2004. "Global Nitrogen: Cycling out of Control." *Environmental Health Perspectives* 112 (10). https://doi.org/10.1289/ehp.112-a556.

"For Plankton, One Size Doesn't Fit All." 2018. *American Museum of Natural History* (blog). February 16, 2018. https://www.amnh.org/explore/news-blogs /news-posts/for-plankton-one-size-doesn-t-fit-all.

Global Invasive Species Database (2023) Species profile: *Mnemiopsis leidyi*. Accessed on December 29, 2023. http://www.iucngisd.org/gisd/species .php?sc=95.

Hitchcock, Charles H. 1873. "Footprints in the Rocks." *Popular Science Monthly* 3 (August). https://en.wikisource.org/wiki/Popular_Science_Monthly/Volume _3/August_1873/Footprints_in_the_Rocks.

Javidpour, Jamileh, Juan Carlos Molinero, Eduardo Ramírez-Romero, Patrick Roberts, and Thomas Larsen. 2020. "Cannibalism Makes Invasive Comb Jelly, Mnemiopsis Leidyi, Resilient to Unfavourable Conditions." *Communications Biology* 3 (1). https://doi.org/10.1038/s42003-020-0940-2.

Losso, Sarah R., and Javier Ortega-Hernández. 2022. "Claspers in the Mid-Cambrian Olenoides Serratus Indicate Horseshoe Crab–like Mating in Trilobites." *Geology* 50 (8): 897–901. https://doi.org/10.1130/g49872.1.

Machemer, Theresa. 2020. "Invasive Comb Jellies Might Overproduce Babies in Summer to Eat Them in Winter." *Smithsonian Magazine*, May 11, 2020. https://www.smithsonianmag.com/smart-news/new-study-suggests-comb -jellies-have-babies-eat-them-winter-180974841/.

Medina, Miles, D. A. Kaplan, Eric C. Milbrandt, Dave Tomasko, Ray Huffaker, and Christine Angelini. 2022. "Nitrogen-Enriched Discharges from a Highly Managed Watershed Intensify Red Tide (Karenia Brevis) Blooms in Southwest Florida." *Science of the Total Environment* 827 (June): 154149. https://doi .org/10.1016/j.scitotenv.2022.154149.

Richardson, Anthony J., Andrew Bakun, Graeme C. Hays, and Mark J. Gibbons. 2009. "The Jellyfish Joyride: Causes, Consequences and Management Responses to a More Gelatinous Future." *Trends in Ecology and Evolution* 24 (6): 312–22. https://doi.org/10.1016/j.tree.2009.01.010.

Sciscio, Lara. 2022. "My Job Is Full of Fossilised Poop, but There's Nothing Icky about Ichnology." *The Conversation*, May 6, 2022. https://theconversation.com /my-job-is-full-of-fossilised-poop-but-theres-nothing-icky-about-ichnology -182906.

Scripps Institution of Oceanography. n.d. "FAQ: Ocean Deoxygenation." Accessed December 29, 2023. https://scripps.ucsd.edu/research/climate-change -resources/faq-ocean-deoxygenation.

Shiganova, Tamara A., Louis Legendre, Alexander S. Kazmin, and Paul Nival. 2014. "Interactions between Invasive Ctenophores in the Black Sea: Assessment of Control Mechanisms Based on Long-Term Observations." *Marine Ecology Progress Series* 507 (July): 111–23. https://doi.org/10.3354/meps10806.

Smithsonian Ocean. 2023a. "Jellyfish and Comb Jellies." https://ocean.si.edu /ocean-life/invertebrates/jellyfish-and-comb-jellies.

———. 2023b. "Sea Walnut (Mnemiopsis Leidyi)." https://ocean.si.edu/ocean -life/invertebrates/sea-walnut-mnemiopsis-leidyi.

Water Science School. 2019. "Dissolved Oxygen and Water." U.S. Geological Survey. October 22, 2019. https://www.usgs.gov/special-topics/water-science -school/science/dissolved-oxygen-and-water.

Great White Shark

Dahlgren, Kristen. 2019. "Great White Shark Spotted in Long Island Sound for First Time Ever." https://www.nbcnews.com/video/great-white-shark-spotted -in-long-island-sound-for-1st-time-ever-60035653662.

Failla, Kathleen Saluk. 1993. "Harbor Seals Returning to the Sound." *New York Times*, February 7, 1993. https://www.nytimes.com/1993/02/07/nyregion /harbor-seals-returning-to-the-sound.html.

NOAA Fisheries. 2023. "Marine Mammal Protection Act: Policies, Guidance, and Regulations." Office of Protected Resources. May 2, 2023. https://www .fisheries.noaa.gov/national/marine-mammal-protection/marine-mammal -protection-act-policies-guidance-and-regulations.

———. n.d. "Atlantic Shark Fisheries Management Highlights: A Timeline." https://www.fisheries.noaa.gov/sustainable-fisheries/atlantic-shark-fisheries -management-highlights-timeline#1997.

———. n.d. "Magnuson Stevens Act." https://www.fisheries.noaa.gov/topic /laws-policies/magnuson-stevens-act.

"Why Are We Seeing More White Sharks and Seals in the Gulf of Maine?" 2020. *New Hampshire Sea Grant* (blog). June 28, 2020. https://seagrant.unh.edu /blog/2020/08/why-are-we-seeing-more-white-sharks-seals-gulf-maine.

Eastern Oyster

Ajmera, Rachael. 2021. "Do Oysters Make You Horny?" Healthline. July 21, 2021. https://www.healthline.com/nutrition/do-oysters-make-you-horny.

Alves, Mickaël Teixeira, Nick Taylor, and Hannah J. Tidbury. 2021. "Understanding Drivers of Wild Oyster Population Persistence." *Scientific Reports* 11 (1). https://doi.org/10.1038/s41598-021-87418-1.

Ault, Alicia. 2017. "Are Oysters an Aphrodisiac?" *Smithsonian Magazine*, February 13, 2017. https://www.smithsonianmag.com/smithsonian-institution/are-oysters-aphrodisiac-180962148/.

Billion Oyster Project. 2023. "2022 Annual Report." https://www.billionoysterproject.org/financials.

"Chimera (Mythology)." 2022. In *Scholarly Community Encyclopedia*. https://encyclopedia.pub/entry/34889.

Diago, Mike. 2020. "The Fight to Bring Oysters Back to the Hudson River." *Hudson Valley Magazine*, March 5, 2020. https://hvmag.com/life-style/news/billion-oyster-project/.

Eaton-Robb, Pat. 2021. "Oystermen Overcome Pollution, Pandemic to Thrive in LI Sound." *AP News*, August 14, 2021. https://apnews.com/article/oyster-farming-dc023c9ec4ac2fe7f96bb212c8b336b4.

Flynn, Dan. 2012. "Oyster-Borne Typhoid Fever Killed 150 in Winter of 1924-25." Food Safety News. March 28, 2012. https://www.foodsafetynews.com/2012/03/editors-note-in-the-winter/.

Le Page, Michael. n.d. "What Is a Chimera?" *New Scientist*. Accessed December 29, 2023. https://www.newscientist.com/question/what-is-a-chimera/.

Maine Department of Marine Resources. 2023. "Historical Maine Fisheries Landings Data." Data set. https://www.maine.gov/dmr/fisheries/commercial/landings-program/historical-data.

Matchar, Emily. 2018. "As Storms Get Bigger, Oyster Reefs Can Help Protect Shorelines." *Smithsonian Magazine*, January 8, 2018. https://www.smithsonianmag.com/innovation/storms-get-bigger-oyster-reefs-can-help-protect-shorelines-180967774/.

"Medieval Bestiary." n.d. David Badke. Accessed December 29, 2023. https://bestiary.ca/index.html.

Nigro, Carmen. 2011. "History on the Half-Shell: The Story of New York City and Its Oysters." *New York Public Library* (blog). June 2, 2011. https://www.nypl .org/blog/2011/06/01/history-half-shell-intertwined-story-new-york-city-and -its-oysters.

Resnick, Brian. 2016. "Behold the Many Wonders of the Oyster, the Sex-Crazed Champion of the Sea." *Vox*, April 6, 2016. https://www.vox.com/2016/4/6 /11352354/oysters-environment-farming.

Spires, Jason E., Elizabeth W. North, Stephanie Westby, Cecily N. Steppe, James M. Dumhart, and Sierra S. Hildebrandt. 2023. "Direct Setting of Eastern Oyster (Crassostrea Virginica) Larvae Confirmed with Calcein, a Fluorochrome Dye." *Estuaries and Coasts* 46 (3): 723–39. https://doi.org/10.1007/s12237-023-01178-x.

"Typhoid Fever in the Large Cities of the United States in 1924." 1925. *Journal (American Water Works Association)* 14 (2): 123–42. https://www.jstor.org /stable/41227050?seq=3. Reprinted by permission from the *Journal of the American Medical Association*, 4:11, March 14, 1925, page 813.

Wachtel, Maddy, and Maureen Dunn. 2020. "Oyster Reef Restoration in New York Relies on Key Ingredient" Interview by the Pew Charitable Trusts. December 21, 2020. https://www.pewtrusts.org/en/research-and-analysis /articles/2020/12/21/oyster-reef-restoration-in-new-york-relies-on-key -ingredient.

Xu, Hong, Ziqiang Zhu, Yiming Huang, Larry D. Bozulic, Lala R. Hussain, Jun Yan, and Suzanne T. Ildstad. 2012. "Innate and Adaptive Immune Responses Are Tolerized in Chimeras Prepared with Nonmyeloablative Conditioning." *Transplantation* 93 (5): 469–76. https://doi.org/10.1097/tp.0b013e318242bddf.

Zraick, Karen. 2021. "11 Million New Oysters in New York Harbor (but None for You to Eat)." *New York Times*, December 10, 2021. https://www.nytimes .com/2021/12/10/nyregion/oysters-new-york-hudson-river.html.

North Atlantic Right Whale

Bishop, Ana L., Leah M. Crowe, Philip K. Hamilton, and Erin Meyer-Gutbrod. 2022. "Maternal Lineage and Habitat Use Patterns Explain Variation in the Fecundity of a Critically Endangered Baleen Whale." *Frontiers in Marine Science* 9 (June). https://doi.org/10.3389/fmars.2022.880910.

Ganster, Rynda, "Identifying Predictive Factors of Terminal Agitation in Hospice Patients: A Retrospective Study" (2018). *West Chester University Doctoral Projects*. 29. https://digitalcommons.wcupa.edu/all_doctoral/29

Gibbens, Sarah. 2021. "Bad Breeding Season Spells Trouble for Endangered Whale." *National Geographic*, May 3, 2021. https://www.nationalgeographic.com /animals/article/north-atlantic-right-whales-calving-season-extinction-spd.

Hamilton, Philip K., and Lindsay A. Cooper. 2010. "Changes in North Atlantic Right Whale (Eubalaena Glacialis) Cow-Calf Association Times and Use of the Calving Ground: 1993-2005." *Marine Mammal Science* 26 (4): 896–916. https:// doi.org/10.1111/j.1748-7692.2010.00378.x.

Knowlton, Amy R., James S. Clark, Philip K. Hamilton, Scott D. Kraus, Heather M. Pettis, Rosalind M. Rolland, and Robert S. Schick. 2022. "Fishing Gear Entanglement Threatens Recovery of Critically Endangered North Atlantic Right Whales." *Conservation Science and Practice* 4 (8). https://doi.org/10.1111 /csp2.12736.

Moura, Paula. 2023a. "To Find Right Whales, Some Scientists Want to Find Their Food's Food." *WBUR*, May 17, 2023. https://www.wbur.org/news /2023/05/17/right-whales-plankton-gas-research.

———. 2023b. "Feds Want Tougher Rules to Protect Whales, Frustrating Boaters." *WBUR*, June 8, 2023. https://www.wbur.org/news/2023/06/08/feds -want-tougher-rules-to-protect-whales-frustrating-boaters.

NOAA Fisheries. 2022a. "North Atlantic Right Whale." February 3, 2022. https://www.fisheries.noaa.gov/species/north-atlantic-right-whale.

———. 2022b. "North Atlantic Right Whale: Conservation and Management." February 3, 2022. https://www.fisheries.noaa.gov/species/north-atlantic-right -whale/conservation-management.

———. 2023a. "2017–2023 North Atlantic Right Whale Unusual Mortality Event." December 8, 2023. https://www.fisheries.noaa.gov/national/marine-life -distress/2017-2023-north-atlantic-right-whale-unusual-mortality-event.

———. 2023b. "North Atlantic Right Whale Calving Season 2024." December 28, 2023. https://www.fisheries.noaa.gov/national/endangered-species -conservation/north-atlantic-right-whale-calving-season-2024.

Ogrysko, Nicole, and Esta Pratt-Kielley. 2023. "Right Whales Aren't Having a Good Year. The Pressure Is on to Save This Hard-to-Track Species." *Maine Public*, April 13, 2023. https://www.mainepublic.org/environment-and -outdoors/2023-04-13/right-whales-arent-having-a-good-year-the-pressure -is-on-to-save-this-hard-to-track-species.

Oldfield, Margery L. 1988. "Threatened Mammals Affected by Human Exploitation of the Female-Offspring Bond." *Conservation Biology* 2 (3): 260–74. https://doi.org/10.1111/j.1523-1739.1988.tb00183.x.

Parks, Susan E., Dana A. Cusano, Sofie M. Van Parijs, and Douglas P. Nowacek. 2019. "Acoustic Crypsis in Communication by North Atlantic Right Whale Mother–Calf Pairs on the Calving Grounds." *Biology Letters* 15 (10): 20190485. https://doi.org/10.1098/rsbl.2019.0485.

Pulver, Dinah Voyles. 2023. "This Critically Endangered Whale Is Trying to Make a Comeback. Humans Keep Killing Them." *USA Today*, October 23, 2023. https://www.usatoday.com/story/news/nation/2023/10/23/so-many-whales-are -dying-that-a-growing-birth-rate-isnt-helping/71233381007/.

Sasson, Anthony. 2020. "Freshwater Mussels in the Midwest." *Midwest Biodiversity Institute* (blog). June 1, 2020. https://midwestbiodiversityinst.org /publications/articles/freshwater-mussels-in-the-midwest.

Savoca, Matthew S., Max F Czapanskiy, Shirel R. Kahane-Rapport, William T. Gough, James A. Fahlbusch, K. C. Bierlich, Paolo S. Segre, et al. 2021. "Baleen Whale Prey Consumption Based on High-Resolution Foraging Measurements." *Nature* 599 (7883): 85–90. https://doi.org/10.1038/s41586-021-03991-5.

Scott, R. 2002. "Eubalaena glacialis (North Atlantic right whale)." Animal Diversity Web. 2002. Accessed December 29, 2023. https://animaldiversity.org /accounts/Eubalaena_glacialis/.

Sommer, Lauren. 2023. "This Fishing Gear Can Help Save Whales. What Will It Take for Fishermen to Use It?" *WBUR*, June 25, 2023. https://www.wbur.org /npr/1183727733/this-fishing-gear-can-help-save-whales-what-will-it-take -for-fishermen-to-use-it.

Tulloch, Vivitskaia J. D., Éva E. Plagányi, Christopher J. Brown, Anthony J. Richardson, and Richard J. Matear. 2019. "Future Recovery of Baleen Whales Is Imperiled by Climate Change." *Global Change Biology* 25 (4): 1263–81. https:// doi.org/10.1111/gcb.14573.

American Horseshoe Crab

Bicknell, Russell D. C., and Stephen Pates. 2020. "Pictorial Atlas of Fossil and Extant Horseshoe Crabs, with Focus on Xiphosurida." *Frontiers in Earth Science* 8 (July). https://doi.org/10.3389/feart.2020.00098.

Cohen, Steve. 2023. "President Biden's Environmental Record." *State of the Planet* (blog). June 5, 2023. https://news.climate.columbia.edu/2023/06/05/president-bidens-environmental-record/.

Earthjustice. 2022. "What the Inflation Reduction Act Means for Climate." *Earthjustice* (blog). August 16, 2022. https://earthjustice.org/brief/2022/what-the-inflation-reduction-act-means-for-climate.

Eisner, Chiara. 2023. "Coastal Biomedical Labs Are Bleeding More Horseshoe Crabs with Little Accountability." *NPR*, June 30, 2023. https://www.npr.org/2023/06/10/1180761446/coastal-biomedical-labs-are-bleeding-more-horseshoe-crabs-with-little-accountabi.

Hayward, Paul. n.d. "Vincent of Beauvais, Speculum Historiale." Lancaster University Department of History. Accessed December 29, 2023. https://www.lancaster.ac.uk/staff/haywardp/hist424/seminars/Beauvais.htm.

Parker, Christine, and John E. O'Reilly. 1991. "Oxygen Depletion in Long Island Sound: A Historical Perspective." *Estuaries* 14 (3): 248. https://doi.org/10.2307/1351660.

Peyer, Suzanne M., Alice J. McCarthy, and Carol Eunmi Lee. 2009. "Zebra Mussels Anchor Byssal Threads Faster and Tighter than Quagga Mussels in Flow." *The Journal of Experimental Biology* 212 (13): 2027–36. https://doi.org/10.1242/jeb.028688.

Popovich, Nadja, Livia Albeck-Ripka, and Kendra Pierre-Louis. 2021. "The Trump Administration Rolled Back More than 100 Environmental Rules. Here's the Full List." *New York Times*, January 21, 2021. https://www.nytimes.com/interactive/2020/climate/trump-environment-rollbacks-list.html.

Reinecke, Antje, Luca Bertinetti, Peter Fratzl, and Matthew J. Harrington. 2016. "Cooperative Behavior of a Sacrificial Bond Network and Elastic Framework in Providing Self-Healing Capacity in Mussel Byssal Threads." *Journal of Structural Biology* 196 (3): 329–39. https://doi.org/10.1016/j.jsb.2016.07.020.

Waldman, John. 1988. "Outdoors; Finding Exotic Fish in Unusual Settings." *New York Times*, September 11, 1988. https://www.nytimes.com/1988/09/11/sports/outdoors-finding-exotic-fish-in-unusual-settings.html.

Whitney, Kristoffer, and Jolie Crunelle. 2023. "Horseshoe Crab Blood Is Vital for Testing Intravenous Drugs, but New Synthetic Alternatives Could Mean Pharma Won't Bleed This Unique Species Dry." *The Conversation*, October 12, 2023.

https://theconversation.com/horseshoe-crab-blood-is-vital-for-testing
-intravenous-drugs-but-new-synthetic-alternatives-could-mean-pharma
-wont-bleed-this-unique-species-dry-214622.

Humpback Whale

Borenstein, Seth. 2021. "Study: Northwest Heat Wave Impossible without Climate Change." *Associated Press*, July 7, 2021. https://apnews.com/article /climate-climate-change-science-environment-and-nature-935be069af34aad4 72074d42097af85e.

Choi, Tae–Young, Tae-Ik Choi, Yu-Ri Lee, Seong-Kyu Choe, and Cheol-Hee Kim. 2021. "Zebrafish as an Animal Model for Biomedical Research." *Experimental & Molecular Medicine* 53 (3): 310–17. https://doi.org/10.1038 /s12276-021-00571-5.

Collins, Dave. 2015. "Whales Return to Long Island Sound after Long Hiatus." *Associated Press*, September 23, 2015. https://apnews.com/general-news-5df4e5 9274b944bc8b1c3c39b6635450.

Fredrikson, Annika. 2015. "Why Have Humpback Whales Returned to Long Island Sound?" *The Christian Science Monitor*, September 24, 2015. https://www .csmonitor.com/Science/2015/0923/Why-have-humpback-whales-returned-to -Long-Island-Sound.

Morgan, Rachael, Mette Helene Finnøen, Henrik Jensen, Christophe Pélabon, and Fredrik Jutfelt. 2020. "Low Potential for Evolutionary Rescue from Climate Change in a Tropical Fish." *Proceedings of the National Academy of Sciences of the United States of America* 117 (52): 33365–72. https://doi.org/10.1073 /pnas.2011419117.

Mottl, Amy. 2019. "Polycystic Kidney Disease (PKD) with Dr. Amy Mottl." Interview by Robert Falk. UNC School of Medicine. https://www.med.unc.edu /medicine/news/chairs-corner/podcast/polycystic-kidney-disease-pkd-with-dr -amy-mottl/.

Outtandy, Priya, Claire Russell, Robert Kleta, and Detlef Böckenhauer. 2018. "Zebrafish as a model for kidney function and disease." *Pediatric Nephrology* 34 (5): 751–62. https://doi.org/10.1007/s00467-018-3921-7.

Philip, Sjoukje, Sarah Kew, Geert Jan Van Oldenborgh, F. S. Anslow, Sonia I. Seneviratne, Robert Vautard, Dim Coumou, et al. 2022. "Rapid Attribution Analysis of the Extraordinary Heat Wave on the Pacific Coast of the US and

Canada in June 2021." *Earth System Dynamics Discussions* 13 (4): 1689–1713. https://doi.org/10.5194/esd-13-1689-2022.

Plunkett, Rachel. 2022. "Where the Humpback Feeds." *Explore Spectacular: Interesting Discoveries From 50 Years of National Marine Sanctuaries* (blog). May 2022. https://sanctuaries.noaa.gov/news/may22/explore-spectacular/where-the -humpback-feeds.html.

Torres, Vicente E., Arlene B. Chapman, Olivier Devuyst, Ron T. Gansevoort, Jared J. Grantham, Eiji Higashihara, Ronald D. Perrone, Holly B. Krasa, John Ouyang, and Frank S. Czerwiec. 2012. "Tolvaptan in Patients with Autosomal Dominant Polycystic Kidney Disease." *The New England Journal of Medicine* 367 (25): 2407–18. https://doi.org/10.1056/nejmoa1205511.

Wolfe, Rob. 2021. "Maine Lobstermen Protest Monhegan-Area Wind Project." *Press Herald*, March 21, 2021. https://www.pressherald.com/2021/03/21/maine -lobstermen-protest-monhegan-area-wind-project/.

Index

ABOUT THE AUTHOR

EIREN CAFFALL is a writer and musician whose work has appeared in *Guernica*, *The Los Angeles Review of Books*, *Al Jazeera*, and *The Rumpus*, and on three record albums. She is the recipient of a Whiting Foundation Creative Nonfiction Grant and a Social Justice News Nexus fellowship at Northwestern University, among other awards. She lives in Chicago with her family. *The Mourner's Bestiary* is her first book.